Principles of Justice and Real-World Climate Politics

Principles of Justice and Real-World Climate Politics

Edited by
Sarah Kenehan and Corey Katz

ROWMAN & LITTLEFIELD
Lanham • Boulder • New York • London

Published by Rowman & Littlefield
An imprint of The Rowman & Littlefield Publishing Group, Inc.
4501 Forbes Boulevard, Suite 200, Lanham, Maryland 20706
www.rowman.com

86-90 Paul Street, London EC2A 4NE

Copyright © 2021 by The Rowman & Littlefield Publishing Group, Inc.

All rights reserved. No part of this book may be reproduced in any form or by any electronic or mechanical means, including information storage and retrieval systems, without written permission from the publisher, except by a reviewer who may quote passages in a review.

British Library Cataloguing in Publication Information Available

Library of Congress Cataloging-in-Publication Data

Names: Kenehan, Sarah, 1979- editor. | Katz, Corey, editor.
Title: Principles of justice and real-world climate politics / edited by
    Sarah Kenehan and Corey Katz.
Description: Lanham : Rowman & Littlefield, [2021] | Includes bibliographical
    references and index.
Identifiers: LCCN 2021036331 (print) | LCCN 2021036332 (ebook) |
    ISBN 9781538162682 (cloth) | ISBN 9781538162705 (paperback) |
    ISBN 9781538162699 (epub)
Subjects: LCSH: Environmental justice. | Climatic changes—Moral and ethical aspects.
Classification: LCC GE220 .P75 2021 (print) | LCC GE220 (ebook) |
    DDC 363.7/0561—dc23
LC record available at https://lccn.loc.gov/2021036331
LC ebook record available at https://lccn.loc.gov/2021036332

Contents

	Introduction *Corey Katz and Sarah Kenehan*	1
1	Integrating Justice in Climate Policy Assessments: Toward a Deliberative Transformation of Feasibility *Dominic Lenzi and Martin Kowarsch*	15
2	Governance toward Goals: Synergies, Equity, Feasibility *Idil Boran and Kenneth Shockley*	35
3	Climate Justice in the Nonideal Circumstances of International Negotiations *Michel Bourban*	59
4	International Law as a Basis for a Feasible Ability-to-Pay Principle *Ewan Kingston*	89
5	Climate Justice, Inherited Benefits, and Status Quo Expectations *Lukas H. Meyer*	115
6	Toward Climate Justice: Making the Polluters Pay for Loss and Damage *Md Fahad Hossain, Danielle Falzon, M. Feisal Rahman, and Saleemul Huq*	149
7	Deficient International Leadership as a Feasibility Constraint: The Case of Multilateral Negotiations on Climate-induced Human Mobility *Jörgen Ödalen and Felicia Wartiainen*	171

8	Feasibility and Justice in Decarbonizing Transitions *Ivo Wallimann-Helmer*	191

Index	211
About the Contributors	221

Introduction

Corey Katz and Sarah Kenehan

The most recent scientific reports on climate change tell us that we are nearing a tipping point. If concentrations of greenhouse gases (GHGs) in the atmosphere are not sufficiently reduced, the world will experience dangerous global warming and all of the consequences that come with it. Despite this looming threat and decades of international negotiations, the world as a whole has had little success in reducing GHG emissions, never mind slowing their continued rise. While the 2015 Paris Agreement does set ambitious goals for minimizing warming, even if nations are able to meet their currently promised contributions toward those goals, those emissions reductions alone will not get us to the targets outlined in the agreement. But neither will those contributions represent anything close to what reasonable principles of distributive justice should require of those nations with the greatest responsibility for creating or contributing to global warming. The political and moral failures that have led to this make-or-break moment are only amplified when one considers the progress that the scholarly work on this issue has made. Since the 1980s, political philosophers and moral theorists (among others) have been exploring the moral complexities of global climate change; indeed, there is a robust and comprehensive literature detailing exactly what morality and distributive justice demand of political institutions, businesses, and individuals in the face of climate change. Yet, in the real world of climate negotiations, politics and policymaking, it is unclear how relevant or helpful these theories have been. The aim of this two-volume series is to help bridge the divide between the work of normative theorists and climate action (or inaction). In particular, contributors explore the relationship between what is "feasible" and theories and principles of climate justice, a relationship that is made even more fraught given the pressing urgency of preventing a climate catastrophe. In the first volume of the series, authors explored the connections (or lack thereof) between normative theorizing, normative principles, and climate action. In this, the second volume of the series, authors tackle the strained relationships between climate politics, principles of justice, and concerns of feasibility.

I. BACKGROUND

Scientists have long been warning of the dangerous consequences of a rapidly warming world: intensified hurricanes, droughts and wildfires, sea-level rise, shifting ranges of diseases and illnesses, decreased biodiversity and environmental resilience, to name a few. They similarly have warned of the turmoil and suffering that was likely to result if warming continues unchecked, including, but not limited to, food and water insecurity, unplanned environmental migration, and the worsening of poverty and violence in places already replete with these difficulties. And, sadly, in many of the countries most vulnerable to these harmful impacts are those who have contributed the least to the problem.

The rising global average surface temperature that is leading to these dangerous changes is the outcome of heightened concentrations of GHGs—for example, carbon dioxide, methane, nitrous oxide—in the atmosphere. These gases are produced by a wide range of human activities, from the burning of fossil fuels for energy at both industrial and consumer levels to overall land use (Boran and Katz 2017). Addressing this problem, then, requires that we find ways to not only reduce the concentration of GHGs in the atmosphere by slowing and eventually stopping such activities but also adapt to the climate changes that are now inevitable.

The global effort to achieve both goals officially began in 1992 when the United Nations Framework Convention on Climate Change (UNFCCC) was adopted at the Rio Earth Summit. The UNFCCC came into force in 1994, which started the international process to build a multilateral climate agreement. The first Conference of the Parties (COP) under the UNFCCC convened in 1995. Negotiations at subsequent COPs led to the Kyoto Protocol, the first treaty to result from the UNFCCC process, which was adopted in 1997. The Kyoto Protocol did not require any climate action on the part of developing countries, was never signed by the United States, and most parties failed to meet the targets it set. Negotiations for a new treaty culminated in the Paris Agreement in 2015, which had a markedly different, bottom-up structure than previous efforts. The primary goal of the Paris Agreement is to limit global warming to below 2°C above preindustrial levels and preferably below 1.5°C (UNFCCC 2015). The agreement is based on a structure of nationally determined contributions (NDCs), whereby each nation is free to determine and submit its commitments, with periodic stocktaking and the submission of new NDCs that are expected to grow in ambition.

After the signing of the Paris Agreement, the IPCC examined the feasibility of reaching the goal of limiting warming to under 1.5°C. In the *Special Report on Global Warming of 1.5°C* (2018), researchers explained that human activities have already caused the world to warm between a range of 0.8°C and 1.2°C. At current rates of warming, we are likely to reach 1.5°C above preindustrial levels sometime between 2030 and 2052 (IPCC 2018, 4). They further reported that the climate-related risks to health, livelihoods, food security, water supply, human security, and economic growth are likely to increase the warmer the world gets, with relative vulnerabilities to such risks increasing disproportionately for already disadvantaged populations (IPCC 2018, 9).

In spite of thirty years of global negotiations and policymaking, global GHG emissions have continued to rise and very little has been committed to adaptation funding. Moreover, the IPCC concludes that if major reductions in emissions are not achieved in ten years' time, the world as a whole will be committed to dangerous levels of warming and all of its consequences (IPCC 2018). Thus, we find ourselves in a moment of urgency with which we are frighteningly unprepared to cope. A survey of the current political landscape could lead one to think that humanity is less suited to confront the causes and effects of climate change than ever before. This is only compounded if we are seeking not only effective reductions in GHG emissions but climate action that responds appropriately to the myriad concerns of justice in this context. To be sure, what little has been accomplished does not even approximate basic principles of climate justice, and there is a great deal of resistance to substantively addressing concerns of justice in global climate agreements.

II. CLIMATE JUSTICE AND GLOBAL CLIMATE NEGOTIATIONS

While the fundamental relation between concentrations of GHGs in the atmosphere and global warming is well understood, achieving effective and fair international coordination to respond to the problem has been far more challenging than might have been expected in the early 1990s. One of the reasons for this is that any attempt to address global climate change raises complex problems of justice. First, many of those communities that are most vulnerable to the risks and harms of climate change have contributed the least to the problem. Second, economic capacity to address the problem is not distributed equally around the globe. Third, political communities and generations have clashing interest claims in relation to the burdens of addressing climate change. These circumstances raise pressing questions about distributive justice and how to coordinate global and intergenerational cooperation (Boran and Katz 2017).

In response to such questions, normative theorists have proposed a number of principles to guide the fair distribution of the burdens associated with reducing emissions, including the polluter pays principle (Gosseries 2004; Meyer & Roser 2010; Shue 1999; Singer, 2002), the ability to pay principle (Shue 1999; Caney 2005), the beneficiary pays principle (Caney 2005; Neumayer 2000; and Shue 1999), and an approach that distributes equal per capita emissions among the global population (Singer 2002). Yet, it is unclear to what extent this theoretical work has influenced policymakers and policy development.

In practice, concerns about justice have sometimes been dismissed completely. Consider the "Brazil Proposal." In the late 1990s, during negotiations leading up to the adoption of the Kyoto Protocol, a group of nations put forth a proposal that identified industrialized nations as the historically responsible agents for climate change and held those agents responsible for the brunt of the costs associated with reducing emissions (UNFCCC 1997). This "Brazil Proposal" asserted a conception of fairness grounded in the polluter pays principle, a principle that is well-defended in the normative literature. Despite its solid normative grounding, the proposal was rejected by industrialized nations. This example—one of many—directs us to consider what role normative principles, theories, and theorists should and can play in policymaking.

It would be mistaken to conclude, however, that the normative work done on the issue should be dismissed as irrelevant. A concern for "equity" has always been central to both the work of normative theorists and the UNFCCC. As a basis for equitable cooperation, the UNFCCC recognizes the need to differentiate responsibility between countries in the global climate regime. The text of the UNFCCC reflects this commitment through a principle of common but differentiated responsibilities and respective capabilities (CBDR-RC) (UNFCCC 1992, Article 3). The Paris Agreement adopts this principle, but while it still calls on developed countries to take the lead in climate action, it puts in place a structure where all parties are to do their part based on their national circumstances (Boran and Katz 2017). Thus, we can understand the work of normative theorists as helping to further explain and justify existing commitments to equity and specifying their implications.

The most public example of the world of normative theorists and climate policy working together is chapter 3 of Working Group III's contribution to the IPCC's Fifth Assessment Report (AR5). The chapter, titled "Social, Economic, and Ethical Concepts and Methods," asks what *should be done* about climate change—a decidedly normative question—and uses this concern as one of the chapter's guiding inquiries (Working Group III 2014, 214). The chapter canvasses some of the most important ideas and concepts that have been developed by justice theorists over the last thirty years and includes philosophers among its lead authors. Clearly, then, work on normative theory is not completely irrelevant to climate policy or negotiations.

III. FEASIBILITY

The nature, scale, scope, and complexity of global climate change, however, tests the relevance and applicability of normative theories and our ability to respond effectively to the problem. Stephen Gardiner has described the phenomenon as a "perfect moral storm." He writes, "Our problem is profoundly global, intergenerational, and theoretical. When these factors come together, they pose a 'perfect moral storm' for ethical action. This casts doubt on the adequacy of our existing institutions, and our moral and political theories" (Gardiner 2010, xii). Similarly, Dale Jamieson explains that because human moral psychology evolved when human beings lived in small, close-knit communities, "it would not be surprising if there were questions relating to anthropogenic climate change about which our everyday morality is flummoxed, silent, or incorrect" (Jamieson 2014, 147). For instance, climate change challenges our understanding of wrongful harm (ibid, 148) and makes it difficult to ascribe fault to individuals (ibid, 152). Both theorists likewise draw attention to the other features of the problem that make it difficult to tackle: the spatial scope of global climate change (ibid, 161; Gardiner 2010, 24), the temporal reach of GHGs and the warming they will cause (Jamieson 2014, 165; Gardiner 2010, 32), and the fact that addressing climate change is a complex international and intergenerational collective action problem (Jamieson 2014, 162; Gardiner 2010, 24).

Jamieson and Gardiner point out the numerous feasibility constraints on achieving any action to address climate change, never mind action that comes close to reflecting the principles of climate justice that normative theorists have worked out. Given this large divide between climate policy and action and normative principles, and in recognition of the urgency of our present situation, normative theorists have started thinking about the feasibility of their proposals. To be sure, we find ourselves at a crossroads: given what is at stake, and the short time that we have to act, is it frivolous to contemplate the demands of climate justice and insist that they be represented in policy documents or action? Put differently, to what extent should those working to better understand or achieve climate justice think about the feasibility of their theories and proposals in a real-world context?

In political philosophy, reflection upon the relationship between what is feasible and normative principles and theories has sometimes taken place via a differentiation between "ideal" and "nonideal theory." Scholars have used the term "nonideal theory" in at least three ways (Valentini 2012). First, holding a conception of the appropriate principles of justice fixed (ideal theory), some take nonideal theory to be about what agents who are motivated to act according to those principles should do when there are other agents who should act but aren't (Feinberg 1973; Sher 1997; Murphy 2003; Schapiro 2003; Cullity 2004, Miller 2011, Stemplowska 2016). In the case of climate change, we might wonder, for example, whether developed country parties that do seem motivated to reduce emissions (countries in the EU) should cut their emissions even more to make up for the failure of other developed countries to do so (Maltais 2014). This discussion is in some sense about feasibility, in the sense that it takes "full compliance" to be infeasible and instead develops normative principles for action in the face of the "noncompliance" of some.

Second, some take nonideal theory to name a particular way of theorizing about justice, namely allowing relatively "sticky" institutional structures (like the international system of sovereign states) or social-scientific regularities (like the human tendency to build family groups with their children) to shape the principles of justice that are developed (Miller 2008; 2013). These elements are sometimes referred to as the "circumstances of justice," since they are taken to shape the ability of the relevant agents to accept and abide by the principles being developed (Rawls 1999a). In that sense, these elements are taken to be deep feasibility constraints not just on political action but on the very act of theorizing about justice itself. Taking these concerns seriously, Rawls worked to develop a theory of both domestic and international justice that gives us a picture of a "realistic utopia" (Rawls 1999b). The idea is that it does not make sense to, for example, theorize about what justice would be in a world without an international system of sovereign states, or spend much time thinking about what justice-improvements such a world might provide. This is because the international system of sovereign states appears to be a very sticky institutional structure.

Third, others take themselves to be nonideal theorists insofar as they reject the need to agree upon or even have a vision of a perfectly just society in order to identify steps that could reduce existing injustice (Sen 2006, 2009; Wiens 2012). We do not need a fully formulated account of justice to serve as an end goal to pursue. Defenders of this view also claim that too much weight is put on identifying the end goal at a high level of abstraction. Unlike the previous understanding of nonideal theory, the issue here is not with how theorists should go about developing theories or principles of justice, but with what those interested in social or political action need in hand in order to work to reduce existing injustice.

This relationship between what is feasible and "action guidance" for what we should do in the "here and now" is a topic of growing interest (Gilabert and Lawford-Smith 2012; Lawford-Smith 2013). Indeed, many contributors to this volume reflect not on normative *theorizing* about justice, but instead on the relevance of normative *principles* for guiding the urgent social and political action needed in the face of the climate crisis.

One point that many contributors highlight is that feasibility constraints can change, which can take years or may happen overnight. Thus, a morally worthwhile goal that has been dubbed infeasible may become a reality once existing constraints shift. But it is often difficult to determine the malleability of a particular constraint or to predict the circumstances under which the constraint would be rendered less influential. Before the global COVID-19 pandemic, for instance, one might not have thought it probable that countries would simultaneously close their borders, shut down schools, and indirectly but predictably slow their national economies by shutting down nonessential services and asking people to stay home from work. Nonetheless, in the presence of the right sort of threat, these actions not only became feasible (at least for a while) but were also deemed a public health necessity. (For more on the connection between the feasibility of responding to COVID-19 and GCC, please see Meyer and de Araujo in volume 1.) Similarly, in the context of global climate change, collective global action became more feasible with the election of American president Joe Biden in 2020 and the electoral success that allowed the Democratic Party to take control of the U.S. Senate, which paved the way for the United States to reenter the Paris Agreement.

Even while acknowledging that feasibility constraints can and do change, it may be that in some cases, we don't have the luxury of waiting it out. In cases of extreme moral urgency, we may have to sacrifice some moral goods in order to secure an even greater moral good. Perhaps climate change is one of these instances: because of the short timeframe that we have to act in order to achieve climate stability, it could be that we have to sacrifice (to some extent, at least) what justice demands in terms of distributing the burdens of climate mitigation and adaptation.

Climate justice theorists have begun to reflect on these questions and pay more attention to the issue of feasibility over the last ten years or so (see, for instance, Posner and Weisbach 2010; Broome 2012; Caney 2014; Roser 2015; Gardiner and Weisbach 2016; Heyward and Roser 2016; Kenehan 2017). In this literature, theorists seek to strike a balance between the demands of justice and what can actually be achieved given the complexity of real-world politics and climate negotiations.

Caney (2014), for instance, provides a reflection on the issue of noncompliance in the climate justice sphere. He begins with the assumption that "it is of paramount importance that humanity avoids dangerous climate change" (Caney 2014, 127–128). He likewise assumes noncompliance with the demands of climate justice ("first-order responsibilities") and argues that we need an account that responds to this inevitability (Caney 2014, 134-135). Such an account defines "second-order responsibilities," which refer "to responsibilities that some have to ensure that agents comply with their first order responsibilities" (Caney 2014, 135). As such, Caney is concerned with feasibility in a certain sense. Given the ongoing failure of many agents to meet their duties of justice in the face of climate change, Caney acknowledges the unlikeliness of agents to comply going forward and constructs a response that takes this real-world noncompliance into account.

Other climate justice theorists have not focused on the issue of noncompliance, but instead on whether normative theories and principles are needed for, or relevant to, climate action at all. Broome (2012) proposes an "efficiency without sacrifice" model for justifying climate action. On his account, by understanding GHGs as externalities, we can look to economics to remedy the problem by setting a price on GHGs and bringing them within the market (Broome 2012, 38). Importantly, Broome argues that removing these inefficiencies will not require anyone to sacrifice their interests and so does not depend on resolving conflicts of interest via principles of distributive justice. This is critical for Broome because it is clear that many nations are unwilling to make sacrifices as part of an international climate treaty. Thus, his theory is taking concerns about feasibility seriously: "Political progress over climate change has ground to almost a halt because governments are unwilling to commit their people to sacrifices. Making sacrifices unnecessary is a way to break the logjam and get the process moving again" (Broome 2012, 38).

Others have argued that traditional principles of climate justice are largely irrelevant at this point of urgency and trying to achieve climate action that meets them fails to take feasibility seriously. Rather than pursuing justice, so understood, our exclusive goal should be reducing global GHG emissions sufficiently to prevent catastrophic climate change. For example, Posner and Weisbach (2010) propose a theory called International Paretianism:

> Any treaty must satisfy what we shall call the principle of International Paretianism: all states must believe themselves better off by their lights as a result of the climate treaty. International Paretianism is not an ethical principle but a pragmatic constraint: in the state system, treaties are not possible unless they have the consent of all states, and states only enter treaties that serve their interests. (Posner and Weisbach 2010, 6)

Posner and Weisbach contend that normative theorists fail to consider a serious feasibility constraint on climate action, namely that no nation will act contrary to its own interests. So regardless of the demands of justice, a treaty that moves us toward climate stability would not be agreed upon unless it takes this constraint seriously. This might mean, for instance, that payments are made to the most prolific emitters such that they have a reason to enter a climate treaty (Posner and Weisbach 2010, 7). Gardiner (Gardiner and Weisbach 2016) has objected to this understanding, arguing, among other things, that this is climate extortion:

> Serious interference with the climate system would shape the lives of very many people around the world, and especially future generations, deeply and pervasively, to the extent of becoming at least a major determinant of their life prospects, and perhaps the dominant factor. Arguably, climate extortion exploits this fact, and this is a central part of what makes it ethically problematic. (Gardiner and Weisbach 2016, 126)

As such, according to Gardiner, concerns of justice must necessarily play a role in our response to climate change. In the previous volume, Weisbach and Gardiner continue their debate regarding the relevance of theories and principles of climate justice. Here they are joined by many others who take reflection upon climate justice and feasibility in multiple new directions.

IV. THEMATIC OVERVIEW

Over the course of this two-volume series, contributors respond to and expand upon the concerns we have reviewed above, offer proposals that consider feasibility, and identify and explore unrecognized feasibility constraints on the pursuit of climate justice. In this volume, authors respond to real-world climate politics and policies, offer proposals and analyses that take concerns of feasibility seriously, and identify immediate justice and feasibility concerns with recent proposals for climate action. In particular, authors look at questions of feasibility as they relate to specific international institutions like the IPCC and UNFCCC, as well widely discussed principles of climate justice, including backward-looking principles like polluter pays and forward-looking principles like ability to pay.

To begin, authors Dominic Lenzi and Martin Kowarsch point out that the IPCC's *Special Report on 1.5°C* included an assessment of feasibility for the first time in any IPCC report, though there was no mention of justice as a factor that can shape the feasibility or desirability of particular models. To ameliorate this issue, they call on climate justice theorists to take a more active role in the IPCC assessment process. Not only would this help to bring any implicit normative judgments being made to the fore, but it would be a step toward building the wider public dialogue that is lacking from current expert-driven debates about the feasibility of climate models. Similarly, Idil Boran and Ken Shockley call for a greater recognition of the importance of dialogue in the face of the climate challenge. They argue for a revised notion of how governance can be created through the existing UNFCCC structure that emphasizes "embedded proceduralism" based on the widely shared values of reciprocity and a concern to promote human flourishing. Like Kowarsch and Lenzi, they argue that dialogue has the power to reshape feasibility constraints toward justice, overcoming what they claim is a false dichotomy between those two notions.

In his contribution, Michel Bourban also explores how to address the tension between justice and feasibility within existing UNFCCC structures, looking specifically at the Nationally Determined Contribution reports that the Paris Agreement requires parties to periodically submit. He argues that parties should be required to include an equity statement regarding their proposed activities based on the UNFCCC treaty norm of common but differentiated responsibilities and respective capabilities. He argues that not only could this transparency promote the feasibility of climate justice, but also explains why enacting this institutional reform is itself feasible.

Next, Ewan Kingston proposes that we address the seeming infeasibility of just climate action by examining different principles of climate justice to see if following some of them may be more feasible than others. He argues that action according to the ability to pay principle, understood as a way of fairly dividing the burdens of the shared goals parties have agreed to via the UNFCCC process, may be more politically feasible than action according to other principles. Lukas Meyer takes a similar approach but instead examines principles of justice that aim to take past emissions into account. He argues that the beneficiary pays principle is more coherent than the polluter pays principle when it comes to past emissions, but he concludes that it is currently politically infeasible for the relevant actors to meet the duties of justice shaped by these backward-looking principles. Md Fahad Hossain, Danielle Falzon, M. Feisal Rahman, and Saleemul Huq, on the other hand, argue that financing loss and damage payments by taxing polluting companies and putting a levy on international flights is not only a feasible step toward enacting a polluter pays approach, but is also more feasible than a proposal that would require an international redistribution of wealth.

Jörgen Ödalen and Felicia Wartiainen examine an under-recognized feasibility constraint on climate action: the difficulty of establishing effective climate leadership. Inefficient or nonexistent leadership on the international level threatens to block any aspirations to realize climate justice, despite philosophers' claims that powerful countries have the moral duty to "go first." They examine both the supply side and the demand side of leadership, thus leading to a better understanding of how leadership deficiencies emerge. In the concluding chapter, Ivo Walliman-Helmer, on the other hand, examines technological and political feasibility constraints surrounding negative emissions technologies (NETs). He argues that there are feasibility constraints that stand in the way of not only pursuing climate justice but also ensuring the fair governance of the NETs that will most certainly be deployed in the pursuit of sufficient emissions abatement.

REFERENCES

Boran, Idil and Corey Katz. 2017. "Climate Change Justice." *Routledge Encyclopedia of Philosophy*. Available at: https://www.rep.routledge.com/articles/thematic/climate-change-justice/v-1/sections/intergenerational-justice

Broome, John. 2012. *Climate Matters: Ethics in a Warming World*. New York: Norton.

Caney, Simon. 2005. "Cosmopolitan Justice, Responsibility, and Global Climate Change." *Leiden Journal of International Law* 18: 747–775.

Caney, Simon. 2014. "Two Kind of Climate Justice: Avoiding Harm and Sharing Burdens." *Journal of Political Philosophy* 22(2): 125–149.

Cullity, Garrett. 2004. *The Moral Demands of Affluence*. Oxford: Oxford University Press.

Feinberg, Joel. 1973. "Duty and Obligation in the Nonideal World." *Journal of Philosophy* 70(9): 263–275.

Gardiner, Stephen. 2011. *A Perfect Moral Storm: The Ethical Tragedy of Climate Change*. Oxford: Oxford University Press.

Gardiner, Stephen and David Weisbach. 2016. *Debating Climate Ethics*. Oxford: Oxford University Press.

Gilabert, Pablo and Holly Lawford-Smith. 2012. "Political Feasibility: A Conceptual Exploration." *Political Studies* 60(4) (December): 809–825.

Gosseries, Axel. 2004. "Historical Emissions and Free-Riding." *Ethical Perspectives* 11(1): 36–60.

Heyward, Clare, and Dominic Roser. 2016. *Climate Justice in a Non-Ideal World*. Oxford: Oxford University Press.

IPCC (Intergovernmental Panel on Climate Change). 2018. *Global Warming of 1.5 °C. An IPCC Special Report on the Impacts of Global Warming of 1.5°C Above Pre-industrial Levels and Related Global Greenhouse Gas Emission Pathways, in the Context of Strengthening the Global Response to the Threat of Climate Change, Sustainable Development, and Efforts to Eradicate Poverty*, edited by Valérie Masson-Delmotte, Panmao Zhai, Hans-Otto Pörtner, Debra Roberts, Jim Skea, Priyadarshi R. Shukla, Anna Pirani et al. In Press.

Jamieson, Dale. 2014. *Reason in a Dark Time: Why the Struggle Against Climate Changed Failed—And What it Means for Our Future*. Oxford: Oxford University Press.

Kenehan, Sarah. 2017. "In the Name of Political Possibility: A New Proposal for Thinking About the Role and Relevance of Historical Greenhouse Gas Emissions."

In *Climate Justice and Historical Emissions*, edited by Lukas. H. Meyer and Pranay Sanklecha, 198–218. Cambridge: Cambridge University Press.

Lawford-Smith, Holly. 2013. "Understanding Political Feasibility." *Journal of Political Philosophy* 21(3) (September): 243–259.

Maltais, Aaron. 2014. "Failing International Climate Politics and the Fairness of Going First." *Political Studies* 62(3): 618–633.

Meyer, Lukas and Dominc Roser. 2010. "Climate Justice and Historical Emissions." *Critical Review of International Social and Political* 13(1): 229–253.

Miller, David. 2008. "Political Philosophy for Earthlings." In *Political Theory: Methods and Approaches*, edited Marc Stears, David Leopold, 29–48. Oxford: Oxford University Press.

Miller, David. 2011. "Taking up the Slack? Responsibility and Justice in Situations of Partial Compliance." In *Responsibility and Distributive Justice*, edited by Zofia Stemplowska and Carl Knight, 230–245. Oxford: Oxford University Press.

Miller, David. 2013. *Justice for Earthlings: Essays in Political Philosophy*. Cambridge: Cambridge University Press.

Murphy, Liam. 2003. *Moral Demands in Nonideal Theory*. Oxford: Oxford University Press.

Neumayer, Eric. 2000. "In Defense of Historical Accountability for Greenhouse Gas Emissions." *Ecological Economics* 33(2): 185–192.

Posner, Eric A. and David Weisbach. 2010. *Climate Change Justice*. Princeton: Princeton University Press.

Rawls, John. 1999a. *A Theory of Justice: Revised Edition*. Cambridge: The Belknap Press of Harvard University Press.

Rawls, John. 1999b. *The Law of Peoples: With, The Idea of Public Reason Revisited*. Cambridge, MA: Harvard University Press.

Roser, Dominic. 2015. "Climate Justice in the Straitjacket of Feasibility." In *The Politics of Sustainability. Philosophical Perspectives,* edited by Dieter Birnbacher and May Thorseth, 71–91. London: Routledge.

Schapiro, Tamar. 2003. "Compliance, Complicity, and the Nature of Nonideal Conditions." *Journal of Philosophy* 100(7): 329–355.

Sen, Amartya. 2006. "What Do We Want from a Theory of Justice?" *Journal of Philosophy* 10(5): 215–238.

Sen, Amartya. 2009. *The Idea of Justice*. Cambridge: Harvard University Press.

Sher, George. 1997. *Approximate Justice: Studies in Non-Ideal Theory*. Lanham: Rowman & Littlefield.

Shue, Henry. 1999. "Global Environment and International Inequality." *International Affairs* 75(3): 531–545.

Singer, Peter. 2002. *One World: The Ethics of Globalization*. New Haven: Yale University Press.

Stemplowska, Zofia. 2016. "Doing More Than One's Fair Share." *Critical Review of International Social and Political Philosophy* 19(5): 591–608.

UNFCCC (United Nations Framework Convention on Climate Change). n.d. "Nationally Determined Contributions (NDCs)." Accessed February 16, 2021.

https://unfccc.int/process-and-meetings/the-paris-agreement/nationally-determined-contributions-ndcs/nationally-determined-contributions-ndcs

UNFCCC (United Nations Framework Convention on Climate Change). 1992. "Article 3." Available at: https://unfccc.int/resource/docs/convkp/conveng.pdf

UNFCCC (United Nations Framework Convention on Climate Change), ad hoc Group on the Berlin Mandate. 1997. "Implementation of the Berlin Mandate: Additional Proposals from Parties." UNFCCC. https://unfccc.int/documents/926

UNFCCC (United Nations Framework Convention on Climate Change). 2015. "Adoption of the Paris Agreement." Decision 1/CP.21. Document FCCC/CP/2015/10/Add.1. Paris.

Valentini, Laura. 2012. Ideal vs. Nonideal Theory: A Conceptual Map. *Philosophy Compass* 7(9): 654–664.

Wiens, David. 2012. "Prescribing Institutions Without Ideal Theory." *Journal of Political Philosophy* 20(1): 45–70.

Working Group III to the Fifth Assessment Report of the Intergovernmental Panel on Climate Change. 2014. *Climate Change 2014: Mitigation of Climate Change.* Cambridge: Cambridge University Press. https://www.ipcc.ch/report/ar5/wg3/social-economic-and-ethical-concepts-and-methods/

Chapter 1

Integrating Justice in Climate Policy Assessments[1]

Toward a Deliberative Transformation of Feasibility

Dominic Lenzi and Martin Kowarsch

The importance of feasibility in climate science and politics has grown considerably since the release of the IPCC's *Special Report on 1.5°C* (IPCC 2018). Previously considered unachievable by many scientists and policy analysts, little scientific knowledge was available to support the 1.5°C target when it emerged as part of the Paris Agreement. Seemingly in response, the *Special Report* included an assessment of feasibility for the first time in any IPCC Report. However, as we will show below, this assessment failed to consider how justice concerns affect feasibility or how public policy preferences interact with scientific and policy discourses. In what follows, we argue that climate justice theorists can help to improve feasibility assessments and public discourse about climate policy by taking a more active role in IPCC assessment processes and in the coproduction of mitigation research. In our view, this approach would help bridge current disciplinary divides separating normative theory, economic efficiency, and judgments of political feasibility.

The concept of feasibility brings together, and often conflates, claims about what is possible given natural scientific laws and current institutions, beliefs, and social norms (Galston 2006; Gilabert and Lawford-Smith 2012). In Gilabert and Lawford-Smith's terminology, feasibility includes "hard" constraints that rule out "political proposals on the grounds that they cannot be implemented in practice," and "soft" constraints, which are comparative and merely set "limits on what people are comparatively more likely to do" (Gilabert and Lawford-Smith 2012, 4). These limits are "neither permanent nor absolute" (ibid.).[2] Unfortunately, this distinction struggles to capture debates about feasibility and climate policy, which typically feature *interrelated* natural and social science findings, along with uncertainty and implicit normative judgments about desirability. For instance, climate policy pathways generated by integrated assessment models combine biophysical Earth system models with socioeconomic models, which include policies such as carbon taxation or emissions trading and the likely behavior of firms and consumers in response to such incentives. As such, many model results are somewhere between hard and soft feasibility, but not reducible to either. Models may also omit normatively desirable political possibilities such as wealth redistribution based on unstated beliefs about feasibility. We will expand on these issues in detail subsequently, but even this brief example reveals a complex relationship between descriptive and normative claims in climate policy.

There are a number of reasons why political feasibility should not be assessed narrowly or reductively. While some experts choose to rely upon opinion surveys or economic models of rationality to understand feasibility (e.g., Wicki, Fesenfeld, and Bernauer 2019; Klenert et al. 2018), these attempts are methodologically questionable. For instance, after the release of the IPCC *Special Report on Renewables* (IPCC 2011), the IPCC was criticized for including a renewable energy scenario developed by Greenpeace that some claimed to be overly optimistic and hence infeasible. Some years later, it turned out that the Greenpeace scenario had significantly underestimated the actual worldwide growth of renewables, with so-called "realistic" scenarios underestimating this growth *by a factor of three*, partly because real consumers were willing to pay more for green energy than economic rationality assumed (Creutzig et al. 2017).[3] Policy preferences can also vary significantly between different communication contexts and over time, a fact that opinion surveys struggle to capture. Indeed, policy preferences are usually more responsive to policy narratives and framing than to policy detail (Kahan et al. 2006). Preferences can also change in response to moral considerations (Schuppert and Seidel 2017). Opinion surveys or economic models poorly reflect these contextual elements. Finally, there is an irreducible element of political judgment in feasibility, a point nicely expressed in Bismarck's famous line that politics is the art of the possible. As Galston puts it, "A thin line separates the visionary from the crank, and no algorithm defines the location of that line" (2006, 545). Yet climate scientists may lack political competence to make such judgments, and have on occasion been accused of naiveté about politics and political communication (e.g., Geden 2018). Given these challenges, in what follows we adopt a broad understanding of feasibility that includes notions of possibility, acceptance, and desirability and that explicitly recognizes how feasibility judgments change over time and across contexts.

In the first section, we highlight some of the problems with the IPCC's assessment of feasibility in its *Special Report*. In light of these shortcomings, we see an excellent opportunity for philosophers to contribute to climate policy. However, as we will argue, this requires moving beyond standard approaches in climate justice, which primarily undertake ex post criticism of normative claims in climate mitigation research.[4] While important, we argue that such approaches are insufficient due to the far-reaching entanglement of facts and values in climate policy assessments. Instead, we propose a *collaborative* role for theorists of climate justice within assessment processes, utilizing a pragmatist conception of how knowledge about desirable and feasible climate mitigation policy can be actively coproduced. Further, we argue that this collaborative role includes supporting deliberation processes about climate policy, which are lacking from current expert-driven debates about feasibility. As we argue, deliberation supported by both climate sciences and climate justice theorists can foster collective learning about the future we want, and how to realistically achieve it.

In the second section, we examine both how the IPCC's feasibility assessment overlooked concerns of justice, and we detail some of the potential contributions of climate justice theorists toward improving the selection and understanding of normative concepts. Going beyond this, we argue in the third section that climate justice theorists should coproduce research on climate mitigation policies, and take part in subsequent feasibility assessments. Finally, we explore the benefits of integrating deliberation into feasibility assessments, and we describe the role of climate justice theorists in this process. We conclude that philosophers can contribute to the identification, clarification, and revision of the normative implications of feasibility in climate assessments, improving and potentially transforming current understandings of feasible and desirable climate policy.

I. FEASIBILITY AND EQUITY IN THE *SPECIAL REPORT ON 1.5°C*, AND THE MISSING CONTRIBUTION OF CLIMATE JUSTICE THEORISTS

The assessment of feasibility in the *Special Report* reveals a clear need to involve theorists of climate justice. The *Special Report* divides feasibility into six dimensions, namely *economic, technological, institutional, sociocultural, environmental/ecological,* and *geophysical* feasibility. For reasons of space, we concentrate here only on sociocultural feasibility, and in particular on the subdimension of equity. The discussion of this subdimension begins promisingly by defining "equity" as fairness across generations. Moreover, this understanding of "equity" was detailed in the IPCC's *Fifth Assessment Report*, which drew upon prominent accounts of climate justice (see IPCC 2018, p. 55). The *Special Report* acknowledges four equity asymmetries that frame a 1.5°C° warmer world, namely (a) that some nations have contributed more and/or benefited more from historic emissions, and therefore bear greater responsibilities; (b) that those most affected by climate impacts tend also to be those least responsible for the problem; (c) that some nations possess greater capacities to contribute and decide upon solutions; and (d) that future disparities in wealth may be created as some are left behind in the transition to a low-carbon economy (IPCC 2018, 55).

However, the *Special Report*'s chapter on feasibility drops all reference to these normative issues, and instead merely presents descriptive evidence of modeling results of mitigation pathways or descriptions of technologies. For example, to ascertain how the equity implications of Solar PV technologies affect their feasibility, the *Special Report* cites a series of descriptive papers that either do not mention the words "equity" or "justice" at all (Geels et al. 2017; IEA 2017a; 2017b) or that do not consider the implications of Solar PV for equity (UNEP 2017). In relation to the equity implications of carbon capture technologies, the *Special Report* cites an empirical paper estimating the geological storage of CO_2 (Alcalde et al. 2018). It is anybody's guess what this tells us about intergenerational equity. Yet far from being exceptional, these examples represent a general strategy of reducing normative considerations to descriptive findings, without providing any rationale for doing so. The equity column of this assessment includes no normative literature whatsoever, yet assessment authors were apparently untroubled about speculating about equity.

This strategy of ignoring normative research may be most problematic in relation to carbon dioxide removal technologies (CDR), which are central to the entire *Special Report*. For instance, the *Summary for Policymakers* finds with high confidence that the 1.5°C target is unreachable without CDR (IPCC 2018, SPM, 14). Nonetheless, no normative literature was considered in the examination of the equity implications of CDR.[5] This is despite evidence that implementing CDR may have severe implications for equity (Fuss et al. 2014; Anderson and Peters 2016) and that framing climate change in terms of equity may actually encourage more ambitious climate policies (Patterson et al. 2018; Klinsky et al. 2017). Further, such a move simply ignores the growing normative literature on CDR (Preston 2013; Heyward 2013; McLaren 2016; Morrow and Svoboda 2016; Shue 2017; 2018; Lenzi 2018), as well as the wider climate justice literature which, as we noted, informed the IPCC's understanding of equity. The failure to draw upon *any* normative literature here is difficult to understand and reinforces the suspicion that many in the climate science community do not regard equity as an important aspect of feasibility.

We can begin to get a sense of what the normative literature can contribute here by considering four common criticisms of appeals to feasibility. First, feasibility can misleadingly present normative statements about acceptability as if they were descriptive facts about possibility, as several critics have argued in relation to Posner and Weisbach's (2010) claim that only a standard of "international paretianism" would be feasible in any international climate treaty (Heyward 2012; Caney 2014; cf. Posner and Weisbach 2010, 117). As Heyward argues, "Nothing stops an argument better than saying a proposal is impossible, but 'impossible' is not equivalent to 'costly'" (2012, 94). The danger of such rhetorical sleight of hand is exacerbated by the seemingly technical connotations of feasibility. Yet as Gardiner says, given that any decarbonization pathway involves some social and economic costs, the question of *how much* disruption to tolerate now and how much to defer is at base a moral question (2011b, 317).

Second, the conflation of descriptive claims for normative claims can be misused to further certain political agendas. As Roser (2015, 73) argues, politicians and lobbyists can manipulate popular beliefs about "feasible" decarbonization policies to their own advantage. Appeals to feasibility may also favor the status quo by limiting the set of policies under consideration. As Schuppert and Seidel argue (2017, 103–4), taking existing preferences as fixed introduces an unjustified bias in favor of the current regime, which may be extremely problematic. As they note, even if abolishing slavery is infeasible in a slave-owning society, this does not mean we are justified in ignoring abolitionist proposals.

Third, appeals to feasibility are often vague about the agents about whom they are predicated, sometimes intentionally so. Roser argues that "two strategies of abusing feasibility considerations for pushing a certain agenda" are failing to specify the agent in question or failing to consider other agents for whom an action would be feasible (2015, 76). As Caney notes in relation to Posner and Weisbach's passive formulation of "is infeasible" (2014, 130), this does not specify the agent in question, but clearly refers to the unprincipled stance of wealthy high-emitting nations. Appeals to feasibility may also fail to specify *when* a policy is deemed feasible. As Roser argues, this falsely implies that feasibility is a static judgment, which "should invite scepticism as to whether [the use of feasibility] is anything else but an attempt to justify the status quo" (2015, 77).

Fourth, appeals to feasibility may illegitimately excuse actors unwilling to fulfill their moral responsibilities. This use of feasibility as an excuse (cf. Galston 2006) is also evident in Posner and Weisbach's argument. As Caney (2014) points out, it is one thing for a state to predict that any international treaty that does not serve the interests of high-emitting states is unlikely to be accepted, but quite another thing for a high-emitting state itself to say this. As Caney says, "Infeasibility here is not a bar. It should be called what it is, namely, 'unwillingness'" (2014, 130). Indeed, Roser claims that appeals to insufficient motivation are "typically simply expressions of human selfishness" rather than genuine constraints (2015, 81).

The approach implicit in each of these four criticisms may be classified as *diagnostic*, insofar as they help to identify normative claims bearing on climate policy, which may be difficult to separate from empirical claims. These diagnostic approaches also clarify the understanding of normative concepts, which may be misunderstood by other disciplines. For example, as Kartha et al. (2018) show, climate scenarios tend to utilize overly narrow interpretations of equity, which inadvertently leads to modeling injustice against future generations and against the inhabitants of less-developed countries.

Climate justice research can also serve a *justificatory* role in finding the most compelling normative reasons to select climate policies. As Swift points out, philosophers are adept at identifying values that are worthy of further consideration (2008, 369), and unlike other disciplines, philosophy is in the unique position to explore normative concepts in their own right, the reasons that support them, and the implications that may follow. An obvious place for such contributions would be concerning the understanding of justice. As Kartha et al. (2018) also show, climate modelers tend to accept the grandfathering of historical emissions as a legitimate baseline, seemingly given its political salience. However, this begs large questions about historical responsibility that have not only dominated debates in climate justice but have been central points of contention in climate negotiations. The diagnostic and justificatory benefits of climate justice research would therefore benefit future assessments of the feasibility and desirability of climate policies.

Nonetheless, we also believe climate justice theorists should play a *collaborative* role, which goes well beyond these more familiar diagnostic and justificatory roles, to codetermine the feasible set of policies under consideration. As we will show in the next two sections, this collaborative role requires philosophers to move beyond ex post criticism of climate mitigation research to take an active role in the coproduction of climate policy assessments.

II. CODETERMINING FEASIBLE CLIMATE POLICIES

Our proposal responds to the currently lamentable situation in which the community of climate justice theorists plays little if any part in determining the choices that form the basis for scientific assessments of climate mitigation, including the selection of normative elements such as the model inputs, criteria, and constraints (Lenzi et al. 2018).

Codetermining the feasible set of climate policies may strike some as beyond the scope of what philosophers can and should do. This idea is tacit in many discussions, yet seldom made explicit. One of the few articulations we are aware of occurs in Rawls's account of nonideal justice, according to which nonideal theory "looks for policies and courses of action that are morally permissible and politically possible as well as likely to be effective" (1999, 89).[6] As commentators have noted, Rawls means to separate the normative analysis of permissibility and the descriptive analysis of feasibility (Simmons 2010; Swift 2008). According to Simmons, the latter requires "the expertise of, e.g., political scientists, economists, and psychologists," but not philosophers (2010, 19). As such, Swift writes that "it is for social science to tell us which states of affairs are feasible and how to achieve them" (2008, 364), while political philosophy plays a purely evaluative role in ranking options within this feasible set. This conception of the role of philosophy relies on the mistaken assumption of clearly distinct facts and values. It is also incorrect both as a statement about social science and as applied to climate science in particular. As we will argue, there is no way to eliminate value claims from social science research, or indeed from much natural science. Yet because climate justice has largely avoided this widely addressed issue in the philosophy of science, the interpretation of values within such research is left to disciplines with little understanding of normative concepts.

While it is rare to find explicit engagement with the relationship between facts and values in literature on climate justice, in the philosophy of science the supposed separation of descriptive from normative research has long been questioned (Putnam 2002), and normative judgments have been shown to be essential components of supposedly descriptive policy-oriented research (Douglas 2009). While an adequate consideration of this topic is far beyond the scope of this chapter, we will nonetheless briefly present several compelling arguments that question the value-free ideal (Kowarsch 2016, Ch. 5). First, the famous argument by Richard Rudner shows that whenever there is uncertainty underlying scientific claims, scientists must make value-laden judgments on the potential social consequences of errors. For instance, when scientists claim that the employment of a particular climate-friendly energy technology will have consequences A and B with a likelihood of L—despite remaining uncertainty about L—this involves an implicit judgment that the social consequences of a potential error would be acceptable or at least relatively unproblematic. Depending upon the direction of the error, this could either underestimate risks or could overestimate risks and spark unwarranted alarm about otherwise promising options. Given the prominence of risk and uncertainty judgments in climate science, this already shows how a normative stance on acceptable social consequences is presupposed before normative analysis begins. Second, cognitive or epistemic values or criteria are influenced by, and can reflect, social, political, or ethical ideals. Indeed, Helen Longino (1996, pp. 41–50) has shown how feminist political philosophy implies different cognitive values. Internal consistency as an epistemic criterion may also build on social values such as order, control, and predictability, while epistemic values such as the simplicity or beauty of a hypothesis are influenced by socially constructed aesthetic values (Douglas 2009, p. 91). Finally, there are "thick" ethical concepts in scientific research that seem to be purely descriptive at first sight but clearly have a strong normative component as well (Putnam 2004). The close interplay between facts and values is essential to the meaning of thick ethical concepts (Kowarsch 2016). Indeed, feasibility is itself an example of a thick concept as are development, sustainability, or planetary boundaries. Each of these implies normative judgments about acceptable benefits and burdens or risks, along with descriptive claims about nature or society.

The analytical lens of fact/value entanglement also allows us to see why philosophers cannot ignore supposedly descriptive aspects of feasibility and desirability. Even if we recognize that normative claims can hide under the seemingly neutral descriptive veneer of feasibility, moral claims can nonetheless easily be mischaracterized as descriptive statements. For example, in some disciplines (especially those influenced by economics), it appears common to think justice can be reduced to the economic costs of mitigation (see Kowarsch 2016, Ch. 8). However, the implicit normative judgments in climate economics are often hard to identify even for those researchers who are actually willing to make them transparent and to discuss them. For instance, some scenarios featuring BECCS (a bioenergy-based CDR technology) that reduce the overall economic costs of climate mitigation are assumed to *satisfy* the demands of intergenerational equity because future generations will pay marginally less (e.g., Kriegler et al. 2013). While these costs certainly matter for intergenerational equity, they are obviously not the whole story. For instance, there are potentially serious injustices resulting from the implementation of BECCS technology that are independent of economic costs, and the very choice to impose CDR reliance upon future generations can treat them as collateral in a climate policy gamble due to the uncertain viability of these technologies at scale (Shue 2017; Lenzi et al. 2018; Fuss et al. 2014). Such questionable yet mostly implicit interpretations of normative concepts are a feature, not a bug, of the disciplinary siloing of normative and critical research from supposedly descriptive climate economics and climate mitigation modeling (Markusson et al. 2020).

The analytical lens of fact/value entanglement therefore requires interdisciplinary work by philosophers in order to codetermine the feasible set of policies, and address problematic assumptions, rather than vainly attempting to set all values aside. In assessment processes, this collaborative role works best when philosophers and researchers from other disciplines jointly explore the interdependence between (1) different climate policy goals, (2) the means to achieve them, and (3) the further consequences of such means. Indeed, this interdependence between means and goals is a hallmark of philosophical pragmatism, particularly in the tradition of John Dewey. From this pragmatist perspective, all scientific hypotheses are inherently normative when they are regarded as a means of solving practical problems (Kowarsch and Edenhofer 2016; Kowarsch 2016). Thus, proposed means of solving a particular problem—along with the goals themselves that are part of the problem analysis—have to be evaluated and potentially revised in light of their practical implications. These implications include identifying the direct effects of a particular policy, its unwanted side effects, and its co-benefits.

While this might appear overly abstract to have much bearing on policy, elements of this approach were in fact already adopted by the IPCC AR5 Working Group III assessment (Edenhofer and Kowarsch 2015), achieving a measure of success in informing international decision-makers and the public about the interdependent implications of global climate policy goals, for example, the 2°C warming target. However, even during the AR5, which appears in hindsight to have been the high-water mark of the involvement of climate justice theorists with the IPCC, there was no normative analysis of climate policy results. While there was a chapter devoted to climate justice (i.e., chapter 3), this was isolated from the rest of the report, and not referred to by other chapters (Kowarsch and Edenhofer 2016). Crucially, the climate justice chapter featured no analysis of climate mitigation policy pathways.

In addition to a joint analysis of modeling outputs, the focus on goals, means, and consequences also requires philosophers and modelers to collaboratively examine how normative assumptions are selected, prioritized, and applied at the beginning of the modeling process. This is far from standard practice. The shared socioeconomic pathways (SSPs), model narratives that serve to assess future policy options in the IPCC (O'Neill et al. 2017), illustrate why such involvement is essential early in an assessment process. The SSPs feature narratives about the future that are used to group assumptions about political, social, and technological choices and constraints, and which are then tested in terms of their feasibility with either the 2°C or 1.5°C target. As a result, the SSPs chart future worlds of varying degrees of justice and imply different political and environmental values (Lenzi 2018). To date, however, climate justice theorists have played no part in the construction of these narratives. One consequence is that some narratives and their associated policies may have been ignored.

Philosophical input may also improve the selection of model assumptions or constraints. Note that integrated models currently include Negishi weights, a technical means to ensure that models do not produce any "additional" global redistribution of wealth (see Kowarsch 2016, Ch. 8). While not explicitly justified, this choice appears again to rely upon unstated beliefs about political feasibility. Yet such a choice ignores much research arguing for *decreasing* inequality on moral, political, economic, and environmental grounds (Sen 1999; Singer 2016; Pogge 2002; Piketty 2014; Stiglitz 2015; Armstrong 2017). If important normative dimensions are not included from the outset, it can be difficult to read them back into modeling outputs.

Given the entanglement of normative judgments with descriptive research about climate policy, and hence in all assessments of feasible climate policies, climate justice theorists should collaborate on the co-production of such research and in resulting policy assessments. This collaborative role would significantly improve the existing debate about desirable and feasible climate policy, providing a method to integrate climate justice within climate policy assessments. In the next section, we explore how deliberation can further improve feasibility assessments, and we identify a role for climate justice theorists in this process.

III. TRANSFORMING FEASIBILITY THROUGH DELIBERATION

The second element of this collaborative role is to support deliberation processes about climate policy, promising to transform existing judgments about what is feasible among researchers, decision-makers, and the broader public. This again goes beyond the standard diagnostic and justificatory roles and requires greater scrutiny of opportunities for public participation and the contribution of normative expertise to facilitate deliberation. Deliberation is the exchange of reasons within a dialogue under conditions of fairness and equality (Dryzek 1990). The basic idea is to accommodate and assess alternative views and to facilitate a reflexive learning process about policy, while catalyzing effects upon public policy discourse more generally. Ireland's use of a citizens' assembly on climate change is perhaps the most prominent example,[7] inspiring contemporary deliberation processes in France, the United Kingdom, Spain, and Germany.

Deliberation can have two major positive effects upon public discourses about the broader feasibility and desirability of climate policies. First, inclusive deliberation processes can increase the democratic legitimacy of climate assessments and climate governance. At present, there is insufficient attention to the literature on deliberation in IPCC assessments. Nonetheless, the one and only deliberative study included in the *Special Report* revealed that two-thirds of participants felt excluded from government decisions about biofuels (Longstaff et al. 2015), a finding likely to hold for many other relevant cases. An emphasis on deliberative processes can help avoid a technocratic climate politics in which the public is a passive agent that does not contribute to decision-making. As noted, the *Special Report* highlights asymmetries of decision-making power. However, as Low and Schäfer (2020) point out, inequalities of power and concerns about legitimacy are also evident in debates about the feasibility of achieving the 2°C or 1.5°C temperature targets and the consequent reliance upon CDR technologies. In the foreground of these debates is a narrow notion of technical feasibility that denotes the ability of a computer model to produce a scenario consistent with a temperature target. Yet seemingly narrow technical feasibility judgments "are proxies for wider worldviews, tied up with different understandings of the freedom of scientific inquiry in policy-driven assessments, the proper relationship between science and policy, and the shape of appropriate science communication" (2020, 2). Integrated assessment modelers claim that decarbonization scenarios provide support for policymakers without being policy-prescriptive or without providing predictions of the future, while critics argue that scenarios exert undue power on the policy debate through the selection of which policies to model, along with the adoption of criteria to measure their benefits and risks (2020, 4–5). One consequence of this power is resulting support for the prolonged viability of a carbon-intensive economy via the heavy reliance on CDR technologies in models (Beck and Mahony 2017). To be sure, policy elites and scientific experts enjoy an asymmetry of information and decision-making power, but they are not always sensitive to public policy preferences. Indeed, as Low and Schäfer note, expert judgment often settles questions of policy choice and model constraints without public participation (2020, 3).[8]

This links to the second major rationale for deliberation, which is to increase the epistemic quality of an assessment of policy alternatives (Kowarsch 2016, Ch. 6). Feasibility assessments have to take into account how knowledge and public opinion is constructed, and the potential for deliberative policy learning to occur. Deliberation goes beyond the recognition that preferences can change and actually *creates the conditions under which such change is more likely*. As empirical research in many countries has shown (Curato et al. 2017), deliberation can substantially affect and improve public policy discourse.

As noted at the outset, it is relatively common among some in the climate policy community to rely upon surveys and opinion polls to gauge feasibility. However, an additional epistemic problem with relying on surveys is that public opinion about policy is greatly determined by perceived ideological allegiance to trusted figures, rather than according to their supposed epistemic merits or technical characteristics (Kahan, Jenkins-Smith, and Braman 2011; Kahan et al. 2006). Once a policy or technology is promoted by a politician or political party, public opinion reacts to this in ways that prior survey results may not capture. Public opinion can change in learning processes about policy alternatives, reinforcing the need for a broader conception of feasibility (see the first section). Taking political preferences as a static, given fact, as the survey approach implies, also blithely ignores the dynamic effects of opportunities for political participation, and especially for deliberation. In fact, the one deliberative study referenced by the *Special Report* found that, in comparison to survey results, deliberation greatly increased support for biofuel policies due to learning effects.

In contrast, the effects of deliberation upon feasibility judgments in particular appear to be remarkable. For instance, Ireland's citizens' assembly brought together ninety-nine members of the public to recommend climate policies. The results stood previous estimations of feasibility on their head: each of the thirteen proposals made by the citizens' assembly received overwhelming support, including a new tax to limit greenhouse gas emissions from agriculture, which all political parties had previously ruled out as clearly politically infeasible. A common charge is that deliberation expects too much rationality or moral virtue from citizens, yet this charge assumes that the status quo exhausts what is politically possible. Instead, as Niemeyer says, deliberation seems to "engender a set of possibilities that are incipient, but not commonly exercised in everyday politics" (2014, 31). Moreover, experiments with mini-publics show that deliberation can reduce existing polarization, while encouraging greater convergence on frames of social justice and the common good, even when undertaken among hostile groups such as climate skeptics (Hobson and Niemeyer 2013). In fact, policy overlap and thus public support for particular climate policies are more likely to be achieved via deliberation (Kowarsch & Edenhofer 2016), despite divergent underlying values and worldviews. Deliberation processes can also promote science communication and instrumental learning when science is involved in deliberation.

IPCC assessments should more extensively review and assess deliberative literature, that is, results of deliberation processes outside the assessment process itself, and also seek to integrate sociological studies on multiple values (Schwartz 1994) in order to represent a wider range of societally relevant normative perspectives. It may even be possible to include deliberative mini-publics within assessment processes, representing different regions according to demographic criteria. One reason for such direct involvement is that individuals' principles and values do not easily translate into specific policies, necessitating an iterative dialogue.

Deliberation should be realized both within IPCC (and other) assessment processes, and within broader societal discourses about climate policy. Within assessments, experts can engage in deliberation processes that are structured along the joint exploration of the implications of policy, as we noted with the example of the IPCC AR5. This approach facilitates a policy-relevant, yet wide deliberative learning process about the feasibility of climate policy. While there are several ways to conceive of and justify such deliberation, we noted in the third section how Dewey's pragmatist philosophy and the resulting idea of jointly exploring the ethically relevant implications of policies establish deep connections between integrative assessment and deliberation. Different actors have different bodies of knowledge and perspectives to contribute, including various values, norms, and their applications to policies. Well-designed deliberation processes can bring these perspectives together and facilitate mutual learning.

Philosophers can work alongside other disciplines on a joint deliberation process in order to facilitate learning within policy assessments (Lenzi et al. 2018). The mutual exchange of arguments and perspectives across disciplines could be greatly improved by learning from the suboptimal integration of climate justice within the AR5 (see Kowarsch and Edenhofer 2016). Many of the points made in the third section of this chapter imply that assessment processes should be deliberately designed as integrative deliberation and learning processes across disciplines and perspectives, instead of a multidisciplinary collection of disconnected inputs. Doing so would help to understand the policy implications of principles of climate justice and to include reflective normative judgments within assessments. Another outcome of philosophers engaging in deliberation processes about climate policy may be a reduction in the so-called "value-action gap." This gap manifests as a result of people not knowing in advance which policy would help to realize their values (Stern et al. 1999). If philosophers put more emphasis on interdisciplinary explorations of climate policies, new avenues for realizing particular values or ethical principles may result, thus helping to reduce the value-action gap.

While deliberation and policy learning within assessments is important in its own right, it is also a means to support wider societal deliberation. Assessments produced along the lines suggested above can be a highly valuable support for societal deliberation processes (Kowarsch et al. 2016). Philosophers could contribute to this in various indirect ways. First, they can promote—and help realize—the idea of a joint deliberative assessment process about climate policy, as previously described. Second, climate justice theorists can examine the potential and limitations of deliberation processes for realizing climate justice, about which little has been said to date. Third, philosophers can directly engage in public deliberations about climate policy, based upon their contributions to improving assessments.

IV. CONCLUSION

In the IPCC's upcoming sixth assessment report, the emphasis on the feasibility of scenarios and even individual policies appears stronger than ever.[9] While some argue that this could be done in a scientifically sound manner by linking different scientific literatures (e.g., Waisman, Coninck, and Rogelj 2019), as we have argued above, there are strong reasons for skepticism. It would be a very unfortunate development if the IPCC presented aggregated global-level statements about the comparative feasibility of policy options based on an ad hoc and methodologically suspect combination of surveys, opinion polls, expert judgment, and model results. This would belie the complexity and normative implications of feasibility, would likely be highly misleading, and in the worst case, it could exclude some policies from consideration altogether without any scientific or normative rationale. Further, as we argued in the second section, an exclusive focus on the aggregated feasibility of model scenarios and pathways neglects the interlinkages of goals and means and their problematic normative implications.

In our view, a best-case scenario would be for future IPCC assessments to entirely reframe how feasibility is understood—or even to abandon this potentially dangerous term altogether—by assessing the overall desirability and related requirements of policy pathways in an integrated manner. Given the challenging entanglement of facts and values, achieving this requires integrating concerns of climate justice into these assessments. All three roles of climate justice—diagnostic, justificatory, and collaborative—are important in this endeavor. Philosophers can extensively contribute to the identification, clarification, and revision of the normative implications of feasibility. Indeed, this may also be desirable for other policy fields, and may refine typical understandings of nonideal justice toward more deliberative and integrated approaches. Our vision for future assessments is to co-produce broad analyses of future policy, and with it to facilitate a deliberative learning process about the overall desirability of climate policies and normative positions. This would transform current understandings of feasibility and public acceptance, thereby also improving ethical and economic understandings of "ideal" climate policy. In doing so, co-produced policy assessments would reveal the breadth of moral, political, and socioeconomic considerations that matter for climate policy, fostering inclusive public debate about a desirable climate future and how best to achieve it.

NOTES

1. Both authors contributed equally to this article. Our research was supported by the RIVET project (grant number: 2020-00202), funded by FORMAS Sweden. We thank Kirsten Meyer, Simon Hollnaicher, Henry Shue, and our anonymous reviewers for comments on an earlier draft.

2. There is also a temporal dimension of feasibility, since future options are highly path-dependent upon what we do now. This has important ethical implications, since climate change is a significantly "front-loaded" problem (cf. Gardiner 2011).

3. Models assume that society (and hence all individuals) always minimize overall costs. See https://www.carbonbrief.org/guest-post-why-solar-keeps-being-underestimated.

4. As we will argue in the third section, "ex post" means undertaking normative criticism only once mitigation research has been produced, rather than contributing to such research at an earlier stage. We take this to be the most common approach of theorists of climate justice.

5. This omission is only clear in the Supplementary Materials for chapter 4. Available at: https://www.ipcc.ch/report/sr15/chapter-4-supplementary-materials/.

6. For Rawls, a situation is nonideal if there is not strict compliance with the principles of political morality, or due to unfavorable external circumstances (see Rawls 1999, 5; 90).

7. See https://2016-2018.citizensassembly.ie/en/How-the-State-can-make-Ireland-a-leader-in-tackling-climate-change/How-the-State-can-make-Ireland-a-leader-in-tackling-climate-change.html (accessed January 29, 2019); https://www.buergerrat.de/aktuelles/buergerraete-weltweit/ (accessed June 11, 2020).

8. This is not to say that expertise is not an essential component of rational policymaking. The point is rather that expert judgments about model assumptions or policy choices reflect judgments of political feasibility as well as technical possibility.

9. See https://www.ipcc.ch/site/assets/uploads/2018/04/040820171122-Doc.-6-SYR_Scoping.pdf (accessed June 15, 2020).

REFERENCES

Alcalde, Juan, Stephanie Flude, Mark Wilkinson, Gareth Johnson, Katriona Edlmann, Clare E. Bond, Vivian Scott, Stuart M. V. Gilfillan, Xènia Ogaya, and R. Stuart Haszeldine. 2018. "Estimating Geological CO 2 Storage Security to Deliver on Climate Mitigation." *Nature Communications* 9(1): 2201.

Anderson, Kevin, and Glen Peters. 2016. "The Trouble with Negative Emissions." *Science* 354(6309): 182–83.

Armstrong, Chris. 2017. *Justice and Natural Resources: An Egalitarian Theory*. Oxford: Oxford University Press.

Beck, Silke, and Martin Mahony. 2017. "The IPCC and the Politics of Anticipation." *Nature Climate Change* 7: 311–13.

Caney, Simon. 2014. "Two Kinds of Climate Justice: Avoiding Harm and Sharing Burdens." *The Journal of Political Philosophy* 22(2): 125–49.

Creutzig, Felix, Peter Agoston, Jan Christoph Goldschmidt, Gunnar Luderer, Gregory Nemet, and Robert C. Pietzcker. 2017. "The Underestimated Potential of Solar Energy to Mitigate Climate Change." *Nature Energy* 2(17140): 1–9.

Curato, Nicole, John S. Dryzek, Selen A. Ercan, Carolyn M. Hendriks, and Simon Niemeyer. 2017. "Twelve Key Findings in Deliberative Democracy Research." *Daedalus* 146(3): 28–38.

Douglas, Heather E. 2009. *Science, Policy, and the Value-Free Ideal*. Pittsburgh: University of Pittsburgh Press.

Dryzek, John. 1990. *Discursive Democracy: Politics, Policy, and Political Science*. Cambridge: Cambridge University Press.

Edenhofer, Ottmar, and Martin Kowarsch. 2015. "Cartography of Pathways: A New Model for Environmental Policy Assessments." *Environmental Science & Policy* 51: 56–64.

Fuss, Sabine, Josep G. Canadell, Glen P. Peters, Massimo Tavoni, Robbie M. Andrew, Philippe Ciais, Robert B. Jackson, et al. 2014. "Betting on Negative Emissions." *Nature Climate Change* 4(10): 850–53.

Galston, William A. 2006. "Political Feasibility: Interests and Power." In *The Oxford Handbook of Public Policy*, edited by Michael Moran, Martin Rein, and Robert E. Goodin, 543–56. Oxford: Oxford University Press.

Gardiner, Stephen M. 2011a. *A Perfect Moral Storm: The Ethical Tragedy of Climate Change*. New York: Oxford University Press.

Gardiner, Stephen. 2011b. "Climate Justice." In *The Oxford Handbook of Climate Change and Society*, edited by John S. Dryzek, Richard B. Norgaard, and David Schlosberg. Oxford: Oxford University Press.

Geden, Oliver. 2018. "Politically Informed Advice for Climate Action." *Nature Geoscience*, May, 1.

Geels, Frank W., Benjamin K. Sovacool, Tim Schwanen, and Steve Sorrell. 2017. "Sociotechnical Transitions for Deep Decarbonization." *Science* 357(6357): 1242–44.

Gilabert, Pablo, and Holly Lawford-Smith. 2012. "Political Feasibility: A Conceptual Exploration." *Political Studies* 60(4): 1–17.

Heyward, Clare. 2012. "Review of Climate Change Justice by Eric A. Posner, David Weisbach." *Carbon & Climate Law Review* 6(1): 92–94.

Heyward, Clare. 2013. "Situating and Abandoning Geoengineering: A Typology of Five Responses to Dangerous Climate Change." *PS: Political Science & Politics* 46(1): 23–27.

Hobson, Kersty, and Simon Niemeyer. 2013. "What Sceptics Believe: The Effects of Information and Deliberation on Climate Change Skepticism." *Public Understanding of Science* 22(4): 396–412.

IEA (International Energy Agency). 2017a. *Getting Wind and Sun onto the Grid: A Manual for Policymakers*. Paris: International Energy Agency.

IEA (International Energy Agency). 2017b. *World Energy Outlook 2017*. Paris: International Energy Agency.

IPCC (Intergovernmental Panel on Climate Change). 2011. *Renewable Energy Sources and Climate Change Mitigation. Special Report of the Intergovernmental Panel on Climate Change*. Cambridge: Cambridge University Press.

IPCC (Intergovernmental Panel on Climate Change). 2018. *Global Warming of 1.5°C. An IPCC Special Report on the Impacts of Global Warming of 1.5°C above Pre-Industrial Levels and Related Global Greenhouse Gas Emission Pathways, in the Context of Strengthening the Global Response to the Threat of Climate Change, Sustainable Development, and Efforts to Eradicate Poverty*. Geneva, Switzerland: World Meteorological Organization.

Kahan, Dan M., Hank Jenkins-Smith, and Donald Braman. 2011. "Cultural Cognition of Scientific Consensus." *Journal of Risk Research* 14(2): 147–74.

Kahan, Dan M., Hank Jenkins-Smith, Donald Braman, and John Gastil. 2006. "Fear of Democracy: A Cultural Evaluation of Sunstein on Risk." *Harvard Law Review* 119(4): 1071–09.

Kartha, Sivan, Tom Athanasiou, Simon Caney, Elizabeth Cripps, Kate Dooley, Navroz K. Dubash, Teng Fei, et al. 2018. "Cascading Biases against Poorer Countries." *Nature Climate Change* 8(5): 348–49.

Klenert, David, Linus Mattauch, Emmanuel Combet, Ottmar Edenhofer, Cameron Hepburn, Ryan Rafaty, and Nicholas Stern. 2018. "Making Carbon Pricing Work for Citizens." *Nature Climate Change* 8(8): 669–77.

Klinsky, S., T. Roberts, S. Huq, C. Okereke, P. Newell, P. Davergne, K. O'Brien, et al. 2017. "Why Equity Is Fundamental in Climate Change Policy Research." *Global Environmental Change* 44: 170–73.

Kowarsch, Martin. 2016. *A Pragmatist Orientation for the Social Sciences in Climate Policy*. Vol. 323. Switzerland: Boston Studies in the Philosophy and History of Science. Springer.

Kowarsch, Martin, and Ottmar Edenhofer. 2016. "Principles or Pathways? Improving the Contribution of Philosophical Ethics to Climate Policy." In *Climate Justice in a Non-Ideal World*, edited by Clare Heyward and Dominic Roser, 296–318. Oxford: Oxford University Press.

Kowarsch, Martin, Jennifer Garard, Pauline Riousset, Dominic Lenzi, Marcel J. Dorsch, Brigitte Knopf, Jan-Albrecht Harrs, and Ottmar Edenhofer. 2016. "Scientific Assessments to Facilitate Deliberative Policy Learning." *Palgrave Communications* 2 (16092).

Kriegler, Elmar, Ottmar Edenhofer, Lena Reuster, Gunnar Luderer, and David Klein. 2013. "Is Atmospheric Carbon Dioxide Removal a Game Changer for Climate Change Mitigation?" *Climatic Change* 118(1): 45–57.

Lenzi, Dominic. 2018. "The Ethics of Negative Emissions." *Global Sustainability* 1(e7): 1–8.

Lenzi, Dominic, William F. Lamb, Jérôme Hilaire, Martin Kowarsch, and Jan C. Minx. 2018. "Don't Deploy Negative Emissions Technologies without Ethical Analysis." *Nature* 561(7723): 303.

Longino, Helen. 1996. "Cognitive and Non-Cognitive Values in Science: Rethinking the Dichotomy". In *Feminism, Science, and the Philosophy of Science*, edited by Lynn Hankinson Nelson and Jack Nelson, 39–58. Dordrecht: Kluwer Academic Publishers.

Longstaff, Holly, David M. Secko, Gabriela Capurro, Patricia Hanney, and Terry McIntyre. 2015. "Fostering Citizen Deliberations on the Social Acceptability of Renewable Fuels Policy: The Case of Advanced Lignocellulosic Biofuels in Canada." *Biomass and Bioenergy* 74: 103–12.

Low, Sean, and Stefan Schäfer. 2020. "Is Bio-Energy Carbon Capture and Storage (BECCS) Feasible? The Contested Authority of Integrated Assessment Modelling." *Energy Research & Social Science* 60(101326): 1–9.

Markusson, Nils, Nazmiye Balta-Ozkan, Jason Chilvers, Peter Healey, David Reiner, and Duncan McLaren. 2020. "Social Science Sequestered." *Frontiers in Climate* 2.

McLaren, Duncan. 2016. "Framing out Justice: The Post-Politics of Climate Engineering Discourses." In *Climate Justice and Geoengineering: Ethics and Policy in the Atmospheric Anthropocene*, edited by Christopher J. Preston. London: Rowman and Littlefield.

Morrow, David R., and Toby Svoboda. 2016. "Geoengineering and Non-Ideal Theory." *Public Affairs Quarterly* 30(1): 83–102.

Niemeyer, Simon. 2014. "A Defence of (Deliberative) Democracy in the Anthropocene." *Ethical Perspectives* 21(1): 15–45.

O'Neill, Brian C., Elmar Kriegler, Kristie L. Ebi, Eric Kemp-Benedict, Keywan Riahi, Dale S. Rothman, Bas J. van Ruijven, et al. 2017. "The Roads Ahead: Narratives for Shared Socioeconomic Pathways Describing World Futures in the 21st Century." *Global Environmental Change* 42: 169–80.

Patterson, James J, Thomas Thaler, Matthew Hoffmann, Sara Hughes, Angela Oels, Eric Chu, Aysem Mert, Dave Huitema, Sarah Burch, and Andy Jordan. 2018. "Political Feasibility of 1.5°C Societal Transformations: The Role of Social Justice." *Current Opinion in Environmental Sustainability*, Sustainability governance and transformation 2018, 31: 1–9.

Piketty, Thomas. 2014. *Capital in the Twenty-First Century*. Translated by Arthur Goldhammer. Cambridge: Harvard University Press.

Pogge, Thomas. 2002. *World Poverty and Human Rights*. Cambridge: Polity Press.

Posner, Eric A., and David Weisbach. 2010. *Climate Change Justice*. Princeton: Princeton University Press.

Preston, Christopher J. 2013. "Ethics and Geoengineering: Reviewing the Moral Issues Raised by Solar Radiation Management and Carbon Dioxide Removal." *Wiley Interdisciplinary Reviews: Climate Change* 4(1): 23–37.

Putnam, Hilary. 2004. *The Collapse of the Fact/Value Dichotomy and Other Essays*. Cambridge: Harvard University Press.

Rawls, John. 1999. *The Law of Peoples, with the Idea of Public Reason Revisited*. Cambridge: Harvard University Press.

Roser, Dominic. 2015. "Climate Justice in the Straitjacket of Feasibility." In *The Politics of Sustainability. Philosophical Perspectives*, edited by Dieter Birnbacher and May Thorseth, 71–91. Abingdon, Oxon; New York: Routledge.

Schuppert, Fabian, and Christian Seidel. 2017. "Feasibility, Normative Heuristics and the Proper Place of Historical Responsibility—a Reply to Ohndorf et al." *Climatic Change* 140(2): 101–7.

Schwartz, Shalom H. 1994. "Are There Universal Aspects in the Structure and Contents of Human Values?" *Journal of Social Issues*.

Sen, Amartya. 1999. *Development as Freedom*. Oxford; New York: Oxford University Press.

Shue, Henry. 2017. "Climate Dreaming: Negative Emissions, Risk Transfer, and Irreversibility." *Journal of Human Rights and the Environment* 8(2): 203–16.

Shue, Henry. 2018. "Mitigation Gambles: Uncertainty, Urgency and the Last Gamble Possible." *Philosophical Transactions of the Royal Society A: Mathematical, Physical and Engineering Sciences* 376(2119): 20170105.

Simmons, A. John. 2010. "Ideal and Nonideal Theory." *Philosophy & Public Affairs* 38(1): 5–36.
Singer, Peter. 2016. *One World Now: The Ethics of Globalization*. Revised third edition. Terry Lecture Series. New Haven: Yale University Press.
Stern, P. C., T. Dietz, T. Abel, G. A. Guagnano, and L. Kalof. 1999. "A Value-Belief-Norm Theory of Support for Social Movements: The Case of Environmentalism." *Human Ecology Review* 6(2): 81–97.
Stiglitz, Joseph E. 2015. *The Great Divide*. New York: W. W. Norton & Company.
Swift, Adam. 2008. "The Value of Philosophy in Nonideal Circumstances." *Social Theory and Practice* 34(3): 363–87.
UNEP (United Nations Environment Programme). 2017. *The Emissions Gap Report 2017*. Nairobi: United Nations Environment Programme.
Waisman, Henri, Heleen De Coninck, and Joeri Rogelj. 2019. "Key Technological Enablers for Ambitious Climate Goals: Insights from the IPCC Special Report on Global Warming of 1.5 °C." *Environmental Research Letters* 14(11): 111001.
Wicki, Michael, Lukas Fesenfeld, and Thomas Bernauer. 2019. "In Search of Politically Feasible Policy-Packages for Sustainable Passenger Transport: Insights from Choice Experiments in China, Germany, and the USA." *Environmental Research Letters* 14(8): 084048.

Chapter 2

Governance toward Goals
Synergies, Equity, Feasibility
Idil Boran and Kenneth Shockley

Too often normative debates on climate change force choices between justice and feasibility. In this chapter, we focus on discussions on equity and implementation in the context of the climate regime in conjunction with Agenda 2030 and the sustainable development goals (SDGs). We argue that dichotomizing equity and feasibility is detrimental for climate change governance, particularly when considered in conjunction with multiple global challenges. The global climate crisis is intertwined with multiple global challenges, including a global biodiversity crisis (Dinerstein et al. 2020), wide-ranging impacts on human health and ecosystem health (Rossati 2017; Watts et al. 2015), poverty, undernutrition, food and water insecurity (Marselle et al. 2019; Turral et al. 2011), sudden onset disaster risk and slow onset effects of climate change and disrupted natural systems (Mal et al. 2018), and environmentally induced forced displacement (Baldwin 2017). Rather than asking whether an international agreement is aligned with antecedent principles of justice or equity, we ask how governance can be strengthened over time and synergies built to address multiple societal issues in an integrated manner. We propose an approach—which we call *governance toward goals*—for a multi-centered, multi-actor, and changing governance landscape. Through this approach, we discuss the complexity and interconnectedness of global goals, and advocate a form of embedded proceduralism for political dialogue in which the values of reciprocity and human flourishing are built into climate governance. While the challenges of implementation toward goals on multiple tracks must be taken seriously, we caution against dichotomizing equity and feasibility.

Our approach helps overcome what is in our eyes a false dichotomy integral to current normative debates on feasibility, and opens channels for research and action for addressing planetary challenges. We highlight that climate justice concerns address a complex web of interdependent considerations. So, feasibility, as we understand it isn't as much about the plausibility of a specific effort to resolve a single problem as it is about juggling multiple interdependent problems, determining what trade-offs are possible, and looking for solutions that allow for mutual gains. To this end, we highlight the indispensable role of political dialogue and identify key principles embedded in substantive values of reciprocity and human flourishing. This approach repositions both equity and feasibility in relation to building synergies across a constellation of actors and multiple treaty processes.

In the philosophical literature on feasibility, justice is theorized as a state of affairs to strive to bring about (Gilabert 2017; 2012). A state of affairs is considered feasible if there is a way to bring it about (Gilabert & Lawford-Smith 2012, 809). Intuitively, feasibility pertains to whether or not a goal or policy proposal can actually work. Political philosophers working on this concept have often noted the ambiguity of the concept (Gilabert & Lawford-Smith 2012), stressing that "feasibility has so far proved to be fairly elusive and under-theorized" (ibid., 246, see also Gilabert 2017). Although publications predating these observations exist (e.g., Barry & Valentini 2009), interest in political feasibility increased considerably in recent years (Southwood 2018). While the details vary, a characteristic of this literature is the conceptualization of political feasibility as a "constraint" imposed by physical, psychological, and sociological conditions that purportedly limit what options are plausible and what are not (Gilabert & Lawford-Smith 2012; Lawford-Smith 2013; Posner & Weisbach 2010; Southwood 2018). Although it has its critics (e.g., Gheaus 2013), feasibility is discussed as a constraint to rule out or to challenge political proposals. While central to sound decision-making, denials of political feasibility, as Gardiner points out, "are notoriously treacherous" (Gardiner and Weisbach 2016, 54). There is a real concern that appeals to feasibility are often little more than attempts at obstruction, without allowing further discussion, social change, and transformation. To overcome this problem, conceptual clarifications and categorizations have been offered (e.g., Gilabert and Lawford-Smith 2012). While the recent focus on clarifying the concept of feasibility and categorizing workable approaches is commendable, these discussions oversimplify the issues nonetheless. Justice tends to be theorized as a state of affairs or a scenario to attain, and feasibility as a constraint that either rules out or ranks scenarios. Theorists distinguish between possibility-, probability-, and cost-based constraints (Southwood 2018) for ranking. They also distinguish between binary and scalar approaches. The binary approach is for ruling out options, according to "hard constraints," and the scalar approach is for ranking multiple options comparatively, through an assessment of "soft constraints" (Gilabert & Lawford-Smith 2012, 813). This approach to feasibility might be operationalizable in limited scale decision-making processes, but it does not capture the formidable complexity of climate change. This is particularly pressing when it is recognized that climate change is not a discrete issue, but part of a nexus of interconnected planetary challenges. The multiplicity of global challenges requires an approach to transformational change that embraces this complexity.

While the Paris Agreement marked a cornerstone in the global climate regime, the world remains markedly off track for achieving the transformations needed for a climate safe and resilient future, raising the question of whether, and how, the goals can be reached. The window to keep global temperature rise at 1.5°C is fast closing. This is further complicated by loss of biodiversity, land and forest degradation, and desertification, and the threat of infectious diseases. Climate change and biodiversity loss are intertwined with impacts on human health, the health of ecosystems, water and food security, malnutrition, and environmentally induced forced migration around the world (Marselle et al. 2019). Global warming is linked to multiple goals, including alleviating poverty and food insecurity, public health, life expectancy, reduction of mortality during childbirth, child mortality, quality education, and gender equality, which are taken up in the SDGs (Kroll et al. 2019; Griggs et al. 2013). Goal setting is inescapably also a question of justice and equity.[1] For climate change, setting the temperature limit to 1.5°C was driven by consideration of justice for communities that are extremely vulnerable to impacts, such as small island states and low-lying coastal regions (Ngwadla 2014). To date, much has been written on justice and equity in the context of the climate change regime.[2] While we do not provide a review of these debates here,[3] we highlight that the debates on climate justice in the normative literature have been largely focused on identifying antecedent principles of justice to guide the negotiations toward a global climate agreement. We look beyond the process of negotiating a treaty and ask how governance can be strengthened *after* the framework of the international treaty or agreement is already in place and interlinked with multiple global goals. We stress that considerations of equity and feasibility in the climate regime are inseparably linked to multiple global goals. A key question is how synergies can be maximized so progress can be made on multiple societal issues in an integrated manner. This requires repositioning the normative discourse on justice and feasibility within a multi-centered, multi-actor, and changing governance landscape with multiple goals. This presents a distinctive set of challenges and opportunities, which we discuss. This should by no means imply that our approach gives a clear-cut answer to questions of feasibility. Rather, we move beyond the analytic distinction of justice and feasibility, and turn the spotlight on interlinkages across multiple goals and the indispensable role of political dialogue in governance toward goals. We highlight guiding principles for political dialogue that heed values of reciprocity and human flourishing specially thought out for our approach. We conclude that our account of governance opens a pathway for integrating the values of procedural justice with the substantive values that should guide just climate policy and that our account does so in a way that avoids obstructionist appeals to overly simple accounts of feasibility.

I. BACKGROUND

The Paris Agreement (UNFCCC 2015) and Agenda 2030, with its 17 SDGs (UN 2015) were both adopted in 2015. The Paris Agreement's flexible architecture confers countries considerable leeway, allowing them to make voluntary pledges based on their national circumstances. It supports non-punitive, dialogue-oriented periodic reviews to ratchet up overall ambition over time toward the agreement's long-term goal of keeping global temperature rise well below 2°C, and preferably close to 1.5°C. The Paris governance architecture also has a wide scope. Built in a spirit of building synergies across a wide range of strategies, and in mutual support with Agenda 2030 and the UN SDGs, the climate regime supports mitigation, adaptation, climate finance, loss and damage, land use, and gender-responsive and sustainable transformation pathways. The Paris Agreement also recognizes that its long-term goals could not be achieved without the engagement and support of actors beyond sovereign states and the multilateral process (Boran 2019). The actions of cities, regional governments, businesses, investors, nongovernmental organizations (NGOs), educational and research institutions, international nongovernmental organizations (INGOs), and other local actors are important. The actions of non-state and subnational actors can strengthen and complement the efforts of central governments or support the goals of the global agreement in the face of government inaction.

Following the adoption of the Paris Agreement, scholarship on global climate politics has seen contending, and even conflicting, responses. Some have emphasized its novelty, while others have seen continuity. Robert Falkner described the new agreement as heralding "the beginning of a new era in international climate politics" (2016, 1108), and Thomas Hale described it as representing "a rare case of multilateral adaptation and innovation in the face of gridlock" (2016, 12). Others have cautioned against overstating novelty, and argued that the distinctive features of the new agreement rebalance and refine the constitutive components of the United Nations Framework Convention on Climate Change inked at the 1992 United Nations Conference on Environment and Development (UNCED), known as the Rio Earth Summit. For example, Sandrine Maljean-Dubois (2016) comments on the principle of common but differentiated responsibilities as having grown in flexibility and adaptability while retaining commitment to the original principle. Regardless, few disagree that the world remains far off track when it comes to meeting the global agreement's goals (Bodansky 2016; Hale 2016; Höhne et al. 2017; van Asselt 2016, 139).

In this period, questions of equity in the climate regime have also been the object of heated debate. Some governance scholars, e.g., Keohane (2016), have argued that the successful adoption of the Paris Agreement put an end to disputes over equity. According to this view, the climate regime after Paris is based on voluntary actions and the process after Paris is all about the effectiveness of implementation, and therefore should be on "focusing on trade-offs rather than making general statements about the importance of equity in climate policy" (ibid.). Keohane's comments were met with sharp criticism and firm re-assertions by many scholars that equity's fundamental place in political analysis and practice did not stop with the adoption of the new agreement (Klinsky et al. 2017; Patterson et al. 2018). This kind of disagreement was hardly new. Rather, it was a new iteration in a long-standing debate that positions concerns with equity and justice as opposing considerations to policy implementation, expediency, and effectiveness—which we will group together as the simple view of "feasibility" that we noted in the introduction. This divide was also reflected in the debate between Gardiner and Weisbach (2016). Throughout the 2000s, a breed of scholars with allegiance to mainstream economics and rational approaches to international relations had highlighted a tension between ethical claims and pragmatic constraints (Posner & Weisbach 2010) and promoted a pragmatic approach, which some labeled as "smart" climate change policy (e.g., Lomborg 2010). While details vary, a shared feature of these approaches consists of ranking social issues or policy proposals, and selecting those that would be more effective, on a given criteria. Supporters of this understanding of feasibility constraints (e.g., Posner & Weisbach 2010) have argued that global climate goals should be set in such a way that can attract agreement and expedite implementation. This exemplifies neatly the view, which we reject, that equity and feasibility are two opposing considerations.

Claims that equity *must* be separated from implementation are clearly unfounded. It is counterproductive, both practically and conceptually, to look to dichotomize equity and feasibility in this way for the simple reason that the governance landscape is far more complex than this rarefied conceptual distinction implies. We sketch the theoretical contours of an approach tuned to governance toward goals, where goals are long-term, multiple, and interlinked. Moreover, the theoretical framework we propose is specially designed for a changing political landscape, where actions by multiple actors are needed toward multiple goals simultaneously. We do not imply that the political landscape of environmental governance has changed, because change is endemic to politics. Rather, the theoretical approach we propose is conceived for a political environment in flux, sensitive to multiple actors, and supporting synergies between multiple policy spaces.

We have seen in this section the need to rethink the outdated belief that we must choose between equity and feasibility. The literature, as much as the actual political landscape, which we have sketched, point toward this need.

To this end, we propose an approach that is:

- conceived for multi-actor political dialogue;
- compatible with goal setting as an approach to governance toward the 1.5°C goal of the Paris Agreement, which we call *governance toward goals*; and
- consistent with the goals of multiple governance tracks, including the Convention of Biological Diversity (CBD), the UN Convention on Combating Desertification (UNCCD), and Agenda 2030 and the UN SDGs.

In what follows we point to changes in the landscape of climate governance that make apparent a need for a new account, one that is able to understand a wider range of constraints and values than those in more traditional conceptualizations of global politics.

II. REPOSITIONING THE OUTLOOK ON GLOBAL CLIMATE GOVERNANCE

In recent decades, scholars of global governance have been increasingly moved to rethink long-held conceptions of the normative space and structure of global politics (e.g., Ruggie 2004; Hale & Held 2011; Held & Maffetone 2016). The traditional theory of international relations reflects interactions and relations between sovereign states, assuming a fundamental separation between the domestic and the international spheres. In this picture, international regimes are defined as formal or informal "modes of institutional cooperation among states" (Ruggie 2004, 501). This conception has been contested in light of increasing recognition that the engagement of non-state actors is integral to world politics (e.g., Ferguson & Mansbach 2004). A growing number of governance scholars describe the global governance landscape on climate change in terms of complex interlinkages between states and various actors including local, non-state, regional, and transboundary organizations and networks (e.g., Hale & Held 2011; Held & Maffetone 2016; Young 2017; Andonova, Betsill & Bulkeley 2009; Wapner 1996).

The Paris Agreement reflects these changing attitudes in climate governance through a flexible architecture based on national contributions, a periodic review process to ramp up ambition, ongoing and innovative forums for political dialogue, and its openness to the support of non-state actors. As John Dryzek remarks: "Global climate governance is changing and with it the role of non-state actors in governance" (2017, 797). It should be noted that the interaction of the UNFCCC with non-state actors is not new. The UNFCCC has historically included observer organizations. But over the years, the role of non-state actors, such as civil society organizations and NGOs, has considerably diversified from being primarily that of observing, holding the process accountable, and putting pressure and influencing state decisions and policies, to also performing tasks that are standardly associated with states (e.g., Wapner 1996; Higgott, Underhill & Bieler 2000), or forming partnership *with* states (Pattberg 2010). While states remain central players, a complex web of interaction and partnerships between states and non-state actors has attracted scholarly interest. This has led some to assert a fundamental change of the landscape of climate governance (e.g., Bulkeley et al. 2014; Hoffmann 2011; Okereke, Bulkeley & Schroeder 2009). Others have argued that this results in fragmentation of governance (e.g., Biermann et al. 2009; Van Asselt 2014), while others have argued that it is a new, more variegated, form of governance (e.g., Hickmann 2017; Bulkeley et al. 2014; Chan, Brandi & Bauer 2016), with discussions of a pluralist, hybrid, or inclusive multilateralism, where the UNFCCC serves a facilitative role (e.g., Bäckstrand et al. 2017; Hale 2016).

It is no longer unusual to describe global climate governance as a complex and dynamic interconnected network among sovereign states and multiple actors operating within, between, and beyond state boundaries. The interconnectedness is complex in that it cannot be represented on a unilinear chart, but is rather multi-layered, multi-noded, and uneven. It is dynamic in that connections are not constant, but in motion. Moreover, the Paris Agreement expressly acknowledged the role of non-state actors (UNFCCC 2015), and institutionalized linkages with the broader landscape of action and engagement through a process called the Global Climate Action Agenda. The Marrakech Partnership for Global Climate Action was launched for boosting non-state actor engagement within the multilateral effort in view of enhancing overall ambition in the pre-2020 period, on multiple tracks, including land use, oceans and coastal zones, water, human settlements, energy, transport, and industry. The Global Climate Action portal (formerly called NAZCA), established before Paris, created a global registry that allowed actors to register their climate action and so also an opportunity for the data community to track these actions and their outputs and to identify gaps.

In addition, the Paris Agreement is linked to the SDGs, recognizing the importance of the well-being of humans and nature. Indeed, one of the SDGs (Goal 13, Climate Action) recognizes the importance of reasonable climate policy as part of sustainable development and views this goal as interdependent with other SDGs, including SDG7 (affordable and clean energy), SDG8 (decent work and economic growth), SDG14 (life below water), SDG15 (life on land), and others. The connections between climate policy and sustainable development are far from static, and new outlooks on connections are regularly being discussed. The 2019 Intergovernmental Panel on Climate Change (IPCC) Report, "Climate Change and Land," for example, provides a detailed discussion of the inextricable links between climate change impacts and land use, ecosystems, and details implications for food, water, and energy. Climate adaptation efforts press against efforts to grow economies, even while generating clean energy is clearly in line with reasonable climate policy. Nonetheless the distribution of clean energy technology requires addressing a range of evolving economic, social, and political concerns. Climate change impacts also threaten energy production. To give one example, the risk of glacier outburst in the Himalayan basins due to melting ice (Langenbrunner 2020) threatens hydroelectric energy production and agriculture in the region with devastating ripple effects on energy accessibility, food security, and health.

Pressing questions arise on how to make progress toward the goals of the Paris Agreement and Agenda 2030, with its 17 SDGs. It may seem as though taking the multiplicity of actors and intersecting issues of climate and sustainable development into account makes feasibility more complicated and intractable. We acknowledge that the challenges are enormous, but we maintain that a more dynamic inclusive governance landscape confers major opportunities. However, thinking about equity and feasibility on a single track, and dichotomously, pitting one against the other, is unproductive. It is not desirable to rank the issues, as Lomborg (2010) suggests, and decide which track is the most feasible in light of considerations of economic or political expediency. Global challenges are inextricably interlinked, and the challenge for the planet is to achieve social and environmental goals simultaneously. Global climate governance should reflect this more complex political landscape, shifting from a near-exclusive focus on nation-states to a form of governance that not only includes non-state actors but also acknowledges and integrates a range of considerations that intersect with the full range of the SDGs. This complexity requires not a ranking of which track is most feasible, but a form of governance that reflects the need for both procedural values that reflect our evolving political landscape and substantive values that reflect the importance of the SDGs. With the pressures of this changing political landscape in mind, in the next section, we propose this new theoretical frame, which we call *governance toward goals*.

III. THE BIGGER PICTURE: GOVERNANCE TOWARD GOALS

Governance toward goals, also described as "governance by goal setting" or "governance through goals," is an approach to governance that is gaining prominence. The framework was developed by Biermann et al. (2017) and Kanie & Biermann (2017) in response to the adoption of the 2030 Agenda for Sustainable Development in 2015. The 2030 Agenda was adopted to recognize and address key global challenges simultaneously in an interconnected manner. Accordingly, the Agenda sets multiple mutually supportive goals in economic, social, and environmental areas to be mutually supportive (Kroll et al. 2019). Moreover, the 17 SDGs of the 203 Agenda and the Paris Agreement are interlinked with SDG13 explicitly promoting the goal of Climate Action.

In the literature on governance by goal setting, scholars explore the distinctive features of this approach. For example, Biermann et al. (2017) separate governance through goals from the international legal system, and they put an emphasis on the role of actors beyond the legal climate regime. They explain how the SDGs were developed through an inclusive and comprehensive process, involving the direct input of many actors at various scales (ibid.). Listing and discussing all the features identified by Biermann et al. 2017 is not within the purview of this chapter. Our approach to governance toward goals differs from theirs, however, in that we do not assert a separation *from* the climate regime, but rather *embrace* the complexity described in the previous section, where the UNFCCC process plays a central role, but is linked to actions and concerns beyond the UNFCCC, including the full range of the 17 SDGs.

A key question is how these multiple goals can be reached in a way that does not compromise one goal at the expense of others. As the goals are multiple and interlinked, both equity and feasibility must be repositioned accordingly. We reject thinking in a unilinear logic of feasibility or within the narrative of a single roadmap. Moreover, questions of equity should be asked not as a single issue, for example, as fair distribution of the burdens and benefits of reducing GHG emissions. Fairness in the sharing the burdens and benefits is important, but because multiple goals are inextricably interlinked, their achievement also depends on minimizing trade-offs—that is, when progress toward one goal curbs progress toward another—and maximizing synergies—that is, when progress toward one goal entails progress toward another goal, also called co-benefits (Nilsson et al. 2018; Kroll et al. 2019). For example, progress made on SDG1 (no poverty) and SDG2 (zero hunger) has co-benefits on all other SDGs (Kroll et al. 2019). But trade-offs can also increase vulnerabilities and create imbalances. For instance, progress on

SDG13 (climate action) and SDG12 (responsible consumption and production) can interfere with SDG1 (no poverty) in developing countries. Synergies can also help record progress in ways that alleviate vulnerabilities where such progress is most needed. For instance, progress on SDG6 (clean water and sanitation) and SDG13 (climate action) can help reduce the spread of waterborne infectious disease, which disproportionately affect communities in developing countries. Likewise, progress on clean energy (SDG7) can boost respiratory health in urban centers.

Recent work in global climate governance provides resources for strengthening actions for multiple goals through the engagement of various actors. In an article titled "Reinvigorating Climate Policy" Chan et al. (2015) discuss the conditions of an effective governance framework for climate change after Paris. Although they do not label it as governance toward goals as such, the governance framework they advance is compatible with our proposed approach. Chan et al. (2015) outline four design principles for climate governance. They argue that governance should be *collaborative*, meaning that it should jointly engage the Secretariat, COP Presidencies, and diverse actors; it should be *comprehensive*, meaning that it should bring together existing sources of data collection into a comprehensive framework to provide a systematic overview of the non-state, local, regional initiatives; it should be *evaluative*, meaning that it should set benchmarks and assess progress; and it should be *catalytic*, meaning that it should generate new relationships between actors, and it should diversify and scale up initiatives (ibid., 469–471). In a subsequent paper, Chan et al. (2019) assessed the governance opportunities and risks associated, among other things, with linking climate change and the SDGs. Based on these papers, we have distilled three key features of governance involving multiple types of actors and interlinked issues. These three elements are not exhaustive, but they do reflect some of the indispensable elements of governance toward goals.

First, governance toward goals must be collaborative. It must support collaboration between Secretariats, COP Presidencies, and multiple actors, including state, non-state, local, regional, and other intergovernmental organizations, and multiple treaty tracks. These types of collaboration create opportunities for "pooling the resources and leveraging efforts" (Chan et al. 2015, 470) of diverse organizations and thus are a key factor for political and practical feasibility. Collaboration between multiple Conventions or multilateral processes is also pivotal. For example, collaborative linkages between the United Nations Framework Convention on Climate Change (UNFCCC) and the CBD are particularly important, a point we further elaborate below.

Second, governance toward goals must be transparent. Tracking and verifying commitments made by governments and non-state, local, and regional actors requires transparent data recording, collecting, and monitoring (Hsu et al. 2016). While we do not have the space to expand on the role, challenges, and progress made on data tracking methodologies here, it is worth highlighting their crucial importance for feasibility. Feasibility cannot be discussed in abstract. Data collection and tracking play a critical role in assessing progress toward goals. Researchers underscore the importance of coordinated data collection (Hsu et al. 2020; Hsu et al. 2019; Chan & Amling 2019; Chan et al. 2018) and methodologies for tracking outcomes that reflect both environmental and social impacts. Compatible indicators across multiple action areas are necessary. Importantly, tracking is also key to understanding inequities. For example, Hsu et al. (2016) note that small and medium businesses that take action on climate change tend to have limited resources, which may hinder declaring their efforts in the global registries (304), thus hindering the visibility of their efforts and their needs. Inclusive methodologies for tracking can help understand successful practices and also identify gaps where more actions are needed and where support can be strengthened. The point here is that feasibility should not be understood in terms of siloed treatments of particular problems, holding climate policy subject to artificial "efficiency blinders" (Gardiner and Weisbach, 2016, p. 254). A better approach to climate policy would reflect the multiple interdependent action areas of climate and development, one that integrates our best evidence-driven understanding of how these different areas are connected with one another through a web of political, social, and ecological considerations. That is precisely the purpose of governance toward goals.

Finally, governance toward goals must support synergies. Some actions toward social and environmental goals are synergistic in that they bring co-benefits (Morita and Matsumoto 2015). For instance, promoting healthy food and agricultural practices, reduction of food waste, sustainable buildings and transport, and natural infrastructures can bring enormous co-benefits for climate change mitigation and adaptation, and for promoting just, climate-resilient, and healthy communities. Climate change and biodiversity loss are "two environmental crises of our times, but approaches that deal with one often neglect the other" (Armarego-Marriott 2020; Dinerstein et al. 2020). Climate change and biodiversity decline are threat multipliers and need to be addressed as interlinked issues (Maes et al. 2013). Harmonizing the processes of various UN organizations and Conventions will be pivotal, for example, between UN Climate Change, CBD, UN Convention for Combating Desertification, Agenda 2030, UN Water, Sendai Framework for Disaster Risk Reduction, and others. There are many barriers to harmonizing and bridging multiple processes, the main challenge being that these are distinct multilateral processes with distinct bureaucracies. Nonetheless, efforts to bridge these distinct processes are already taking place. One example is the Interagency Liaison Group (ILG), co-chaired by the Secretariat of the

CBD and the World Health Organization (WHO), with the participation of multiple UN agencies as institutional core members, including the UNFCCC (Jacquemont and Caparrós 2002).

Taking seriously the importance of the interwoven goals of Sustainable Development and Climate Change governance requires rethinking feasibility constraints. As we have considered above, the complex interdependence of the various goals of climate governance requires considering how working toward various goals might influence, constrain, or provide co-benefits with respect to other goals. Governance toward goals takes this interdependence seriously. We will see below how taking this interdependence seriously entails integrating procedural and substantive values into climate governance and that it does so primarily by acknowledging the importance of political dialogue toward the realization of these interdependent goals.

IV. INCLUSIVE POLITICAL DIALOGUE: PROCEDURAL NORMS AND SUBSTANTIVE VALUES

While goal-oriented governance cannot function without political dialogue, dialogue alone does not help resolve the challenges of implementation, and it does not singularly help attain shared goals. Discussion and dialogue can drive people apart, reinforce differences, and irretrievably deepen divisiveness. Dialogue alone does not address issues of equity, and dialogue alone does not make the world more just. But making headway toward shared goals is not possible without political dialogue. Heeding the importance of inclusive dialogue in the pre-2020 framework, in 2018, the COP Presidencies of Fiji and Poland opened a facilitative discursive process known as the "Talanoa Dialogue" in order to create a constructive, non-confrontational, solutions-oriented environment in the Pacific tradition. This initiative was one of the principal features of the pre-2020 efforts to bring together governments with businesses, local and regional authorities, investors, civil society, and other nongovernmental actors. Moving forward, strengthening the channels of political dialogue will be crucial, and they will need to be strengthened specifically for a goal-oriented integrated governance linking climate change with the full suite of SDGs. To this end, political dialogue needs to expressly support collaboration, transparency, and synergies outlined above.

A primary question is how to balance procedures and substantive values in political dialogue. Procedures are the rules and norms to which participants are expected to adhere. Substantive values represent shared visions. Rather than thinking about political dialogue in entirely substantive terms, or in entirely procedural terms, our proposed hybrid approach is conceived for governance toward goals. We call this approach "embedded proceduralism" to distinguish it from "pure" or "strict" forms of proceduralism. The idea is to embed proceduralism within substantive collective goals, thus reflecting the view that there is no hard distinction between procedural and substantive approaches. Our proposed hybrid approach—embedded proceduralism—is less an integration of two distinct approaches and more of an acknowledgment that a hard distinction between these approaches was artificial.

A hybrid approach is appropriate for complex non-unilinear systems with multiple goals. It allows flexibility through procedures of non-confrontational dialogue, while promoting overriding substantive values that support shared goals. Procedure flexibility is important for adaptability to multiple goals and actors, and creates the conditions for political dialogue that embraces pluralism and the inclusion of multiple actors. Substantive values are also important for creating a cohesiveness with the goals; for instance a shared vision of living in harmony with nature was adopted by the CBD, and can be linked to SDG14 (life below water) and SDG 15 (life on land).

Both procedures and substantive values are relevant to equity in the climate regime. On the one hand, when framed as a primarily *substantive* question, equity can be invoked to determine an allocation of roles or responsibilities toward specific goals. Equity can also be invoked to reverse imbalances. There are long-standing debates among scholars of democratic deliberation on how to address competing substantive claims (Gutmann & Thompson 2004). Exclusively procedural values cannot resolve all substantive disagreements, such as balancing the need for economic growth, the need to preserve resources, and the need to reduce energy consumption as a part of an emissions mitigation strategy. On the other hand, procedural principles "do not have to claim to be capable of transcending all fundamental moral disagreements," but they offer "a way of adjudicating the disagreements [in political processes]" (ibid., 130). Typically, procedural approaches presume that all parties start off with recognition of a shared problem they are committed to resolve collectively (Gutmann & Thompson 2004). A hybrid approach, such as ours, that integrates both substantive and procedural values is more suited to the complex forms taken by "equity." Creating an environment for political dialogue in which actors and initiatives are given a voice and representation is an important requirement of equity. Giving voice requires more than just generating a procedure that allows for alternative voices, it requires that there be a shared recognition of the substance of what is claimed by those voices. We must not only include voices but also understand that to include those voices is, in part, to recognize a common concern over shared problems. Both procedural values and substantive values are necessary for a suitable account of equity. For example, local practices, including traditional farming practices, are not always labeled as climate action but nevertheless promote sustainable practices and can be important for adaptation and resilience.

Understanding these local practices involves not just providing a procedure for them to participate in political dialogue, but building into that procedure the means of recognizing the legitimacy of those practices as part of a robust form of climate action. More generally, embedding shared goals within the procedural rules can avoid the pitfalls of the long-standing and too-often unproductive dichotomy that "poses a choice between basing politics on a comprehensive conception of the good, on the one hand, or limiting politics to a conception of procedural justice, on the other" (ibid., 91). Such an artificial separation is not conducive to addressing complex problems; a hybrid approach that integrates substantive values into the procedural guidelines surrounding political dialogue is more reflective of the real challenges facing global climate governance.

As we have explained in this section, a suitably constructed set of procedural guidelines that integrate substantive values can promote collaborative, transparent, and synergistic features of governance toward goals. Political dialogue can be attuned to the national circumstances of state actors developing their nationally determined contributions (NDCs). But inclusive political dialogue must give voice to varying needs and experiences of local, regional, and non-state actors, including, for example, sharing of best practices, successes, and challenges in adaptation through the use of local knowledge or nature-based solutions fitted to maximize the resilience of the local ecosystem to climate impacts. Political dialogue will be necessary for forming the ties that bind together disparate actors along shared norms and principles and toward helping local actions link to regional and intergovernmental processes. In what follows we will propose two guiding principles for governance toward goals, principles that we maintain express the substantive and procedural values so central to the sort of inclusive political dialogue we described above.

V. TWO GUIDING PRINCIPLES

We suggest two guiding principles that support equity through the integration of suitably shared substantive values into fair procedural norms. These guiding principles would be suitable for multiple forums including, but not limited to, periodic global stocktakes, Regional Climate Weeks, the Global Climate Action Agenda (UNFCCC), the Action Agenda for Nature and People (CBD), etc. To the extent that the goals of climate governance have been set, substantive values that are aligned with *broader collective goals* would strengthen and complement proceduralism. For example, the long-term collective goals of the Paris Agreement include first and foremost the goal of limiting of global temperature rise well below 2°C, and making every effort to keep it at 1.5°C. They also include a commitment to enhance the prospects and the conditions of human fulfillment and flourishing, and to take action for resilience and adaptation. Political dialogue[4] must therefore be consistent with the near-term and long-term goals of the Paris Agreement and the other SDGs.

While additional principles may be developed, two guiding principles are:

(1) Principle of Reciprocity: *Parties and other participants in the global climate regime owe each other justification for the proportionality of efforts that they put into the global endeavor.*

The Paris Agreement acknowledges the different circumstances of different nations and the need to work together to increase ambition. The principle of reciprocity reflects this need. According to this principle, parties can be asked to provide an account of how their NDCs satisfy the requirements of reciprocity. For example, parties that discount their mitigation actions due to national circumstances could be asked to provide both a justification of how they benefit from the mitigation actions of other parties as well as an explanation of how this discounting helps them reach certain goals for the next assessment and review process. This kind of justification can help enhance and align transparency standards across parties. In the implementation of the Paris Agreement, countries are the primary actors, but the principle of reciprocity can also be expanded to non-state and subnational actors, whose roles are acknowledged as previously discussed. The principle of reciprocity gives support to create political dialogue where local, regional, and non-state actors can voice their needs and justify their contributions toward goals. Thus, reciprocity highlights an important normative dimension of governance toward goals.

(2) Principle of Human Flourishing: *Parties and other participants owe justification in terms of the effects of their decisions on development, vulnerable populations, and human flourishing more generally.*

Our conceptualization of human flourishing is rooted in the capabilities approach to human flourishing (Sen 1999; Nussbaum 2011). The capabilities approach provides a means of focusing on human flourishing that relies not on any particular state of affairs or condition, but rather on the freedoms and opportunities necessary for a complete life. In the words of Crocker and Robeyns:

> The capability approach . . . asks whether people are able to be healthy, and whether the means or resources necessary for this capability, such as clean water,

adequate sanitation, access to doctors, protection from infections and diseases, and basic knowledge on health issues, are present. It asks whether people are well-nourished, and whether the conditions for the realization of this capability, such as having sufficient food supplies and food entitlements, are being met. It asks whether people have access to a high-quality educational system, to real political participation, and to community activities that support them, that enable them to cope with struggles in daily life, and that foster real friendships. (2009, 64)

The capabilities approach is helpful for focusing on the needs of humans to flourish in times of instability (Shockley 2014), without being limited to times of instability alone. The capabilities approach recognizes that human well-being and flourishing are contextually shaped, confer a broad range of flexibility, and thereby avoid a one-size-fits-all view of well-being. The focus on human needs can therefore be adapted to acutely challenging times of instability and crisis.

The values of reciprocity and human flourishing capture two vital features of the climate regime after the Paris Agreement, namely, the intersubjective, discursive model of presenting, evaluating, reviewing, and revising the NDC submissions, and the need to make climate commitments responsive to the locally determined development needs of communities, including non-state, local, and regional actors, so as to ensure the conditions of human flourishing. As there is not sufficient space here to expand on these principles fully, we have characterized the shape of these principles in broad strokes. By being responsive to substantive values, these principles allow for a more nuanced and contextual approach to equity in climate change governance. As climate change affects different communities differently, and as it amplifies vulnerabilities, focusing on the diverse forms of human flourishing and requiring the sort of contextualist justification in our principle of reciprocity should promote a more equitable climate regime. This drive toward a more contextual and development-sensitive approach to climate governance also speaks to concerns over equity found in the work of scholars, such as Shue (1999, 2010), Jamieson (2014), and Moellendorf (2014), who have highlighted the importance of integrating concerns for development and reduction of poverty into climate change policy. As anticipated above, the 2030 Agenda and the SDGs provide opportunities to respond to these concerns. However, our responses cannot be limited to climate change and poverty reduction, and must include a wider range of concerns, including, for example, adaptation to the health impacts of climate change and biodiversity loss, and living in harmony with nature that is inclusive of traditional and local knowledge and responsive to gender equality.

In this section, we presented two guiding principles that we believe would not only support a more appropriate account of equity in climate governance but that are also in line with our preferred governance toward goals. While further research is needed to develop the operational details, we have explained the importance of integrating reciprocity and human flourishing in the guiding principles of political dialogue. Reciprocity and human flourishing are mutually supportive and complementary. The recognition of voice is central to our understanding of inclusive dialogue, and, as we have described above, requires the procedural features allowing for the representation of those whose voices might not be otherwise heard to be complemented with an acknowledgement of the substantive values presented by those voices. We suggest that, along with reciprocity, human flourishing be a centerpiece of discussions over NDCs: parties should ask of one another to what extent does this NDC contribute to the promotion of human flourishing at the global level, and to what extent does it contribute to or affect human flourishing domestically? Likewise, facilitative dialogues with non-state, local, and regional actors, can also heed this principle. While there are operationalization issues with the SDGs, recognizing human flourishing constitutes the foundation for appeals to equity: the instability and suffering that is already being felt, and likely to intensify due to ever-increasing climate change, is not felt equitably (Shockley 2014, 2018; Moellendorf 2014). In order to support the realization of multiple goals, the scope of dialogues should be expanded to include multiple targets, for example, sustainable land use, food and agricultural practices that support climate mitigation and adaptation goals, re-greening and building natural infrastructure in urban centers, affordable and clean energy, and increased protection against disaster risks.

VI. CONCLUSION

For some time, discussions on equity in the climate regime revolved around substantive accounts of burden allocation in an international setting. Too often, it has been suggested that feasibility is in conflict with equity. Moving away from this dichotomization, this chapter explained that the changing landscape of global climate governance requires a more nuanced approach. We explored governance toward goals, which recognizes the role of governments, local and regional actors, businesses, and civil society to respond to climate change alongside the SDGs. We then highlighted the crucial role of political dialogue, identifying key principles embedded in substantive values of reciprocity and human flourishing.

We distanced ourselves from strict forms of pure proceduralism and developed "embedded proceduralism." Embedded proceduralism allows for both the fairness of proceduralism and substantive values to guide political dialogue in governance toward goals. Procedural principles, we have argued, must be anchored in suitably framed substantive considerations in order to be consistent with the long-term goals of the global climate effort and the SDGs. We discussed how these commitments can be embedded in the procedural principles that guide the political process under the UNFCCC as it moves forward after Paris, and we connected these efforts to the 17 SDGs of Agenda 2030. But, most importantly, the principles we propose support an approach that encompasses equity in a changing landscape of global governance while avoiding what is in our eyes a false dichotomy between equity and feasibility.

Governance toward goals should by no means imply a grand solution to either the considerations of equity or feasibility. It is intended to provide a framework to understand the complex and interconnected global governance landscape toward environmental and social goals. Going back to the old dichotomies that force choices between equity or implementation effectiveness will not help; we will simply be repeating past mistakes and false starts. Those most affected by the impacts of a changing climate and multiple stressors on natural and social systems deserve a more nuanced conception of equity. Those putting their efforts toward a sustainable, equitable, and climate-safe future deserve recognition of their actions through more nuanced conceptions of feasibility.[5]

NOTES

1. A note on terminology: we acknowledge that justice cannot easily be captured by a set definition, as no theorist or school of thought throughout history was able to lay an exclusive claim to how justice should be conceptualized and practiced. There is a vast discourse on justice, with a rich history, spanning across the normative literature and social movements. In this paper, we do not engage this broad literature. Instead, we refer to a particular subfield that forms what could be called the "climate change justice" literature, as reviewed in Boran 2018; Boran & Katz 2017. This subfield is relatively recent, and has developed in response to the proceedings of the global climate regime (Boran 2019, chapter 1). While the details of the perspectives vary, a central concern is that although the effects of climate change are pervasive, those most impacted are those least responsible. Another key concern is the persistent imbalance in the voices and engagement of those who stand to lose as a result of climate and environmental change. We note that the normative literature is limited, and heavily anchored in the subdiscipline of moral and political theory in the English language developed in the last sixty years, where a vast imbalance of representation persists. In the climate policy discourse in the UNFCCC, the term that is favored to discuss similar concerns detailed by philosophers is "equity." Because we do not write exclusively for an audience of normative theorists, but also for a broader interdisciplinary audience that includes policy and practice beyond academia, we adopt an inclusive approach to terminology and use "justice" and "equity" interchangeably. Nor are we alone; in some of the specialized normative literature on climate change, "justice" and "equity"

are used interchangeably (e.g., Gardiner and Weisbach 2016). As to the broader interdisciplinary literature anchored on the UNFCCC, the terms are frequently used interchangeably (e.g., Ngwadla 2014; Patterson 2018). Though we acknowledge that justice and equity are not fully synonymous, we accept some degree of interchangeability by convention.

2. An international regime can be defined as the principles, norms, rules, and decision-making processes in a given area of global governance (Krasner 1983; Young 1989). International regimes include Conventions, such as the UNFCCC, the CBD, the UNCCD. A Convention is an international legal instrument, elaborating on its rules and regulations and may contain the elements of what is to be negotiated over time multilaterally. If a Convention is negotiated in anticipation of further elaboration of texts and agreements, it is called a Framework Convention (Boran 2019, 10, note 1). Throughout this chapter, we refer to international regimes in this context.

3. For reviews, see Boran 2019, chapter 1; Boran 2018; Boran and Katz 2017; Boran & Shockley 2015.

4. Principles of political dialogue for the intergovernmental negotiation process are discussed in Boran 2017. Here, we expand the scope of political dialogue to include a multiplicity of actors and interlinked goals.

5. Acknowledgements: Idil Boran's work on climate change has received support from the Faculty of Liberal Arts and Professional Studies, by the Dahdaleh Institute for Global Health Research, at York University. Kenneth Shockley's work on climate change has been supported by the School of Global Environmental Sustainability and by the College of Liberal Arts, both at Colorado State University. Boran thanks Sander Chan for research collaboration and valuable insights on strengthening climate governance, which informed the discussion in parts of this chapter.

REFERENCES

Andonova, Liliana B., Michele M. Betsill, and Harriet Bulkeley. 2009. "Transnational Climate Governance." *Global Environmental Politics* 9(2): 52–73.

Armarego-Marriott, Tegan. 2020. "Climate or Biodiversity?" *Nature Climate Change* 10(5): 385–385.

Bäckstrand, Karin, Jonathan W. Kuyper, Björn-Ola Linnér, and Eva Lövbrand. 2017. "Non-State Actors in Global Climate Governance: From Copenhagen to Paris and Beyond." *Environmental Politics* 26(4): 561–79.

Baldwin, Andrew. 2017. "Climate Change, Migration, and the Crisis of Humanism: Climate Change, Migration and Humanism." *Wiley Interdisciplinary Reviews: Climate Change* 8(3): e460.

Barry, Christian, and Laura Valentini. 2009. "Egalitarian Challenges to Global Egalitarianism: A Critique." *Review of International Studies* 35(3): 485–512.

Biermann, Frank, Norichika Kanie, and Rakhyun E Kim. 2017. "Global Governance by Goal-Setting: The Novel Approach of the UN Sustainable Development Goals." *Current Opinion in Environmental Sustainability* 26–27(June): 26–31.

Biermann, Frank, Philipp Pattberg, Harro van Asselt, and Fariborz Zelli. 2009. "The Fragmentation of Global Governance Architectures: A Framework for Analysis." *Global Environmental Politics* 9(4): 14–40.

Bodansky, Daniel. 2016. "The Paris Climate Change Agreement: A New Hope?" *The American Journal of International Law* 110(2): 288–319.

Boran, Idil. 2017. "Principles of Public Reason in the UNFCCC: Rethinking the Equity Framework." *Science and Engineering Ethics* 23(5): 1253–71.

Boran, Idil. 2018. "On Inquiry into Climate Justice." In *Routledge Handbook on Climate Justice*, edited by Tahseen Jafry, 26–41. London: Routledge.

Boran, Idil. 2019. *Political Theory and Global Climate Action: Recasting the Public Sphere*. Routledge Focus on Philosophy. London; New York: Routledge, Taylor & Francis Group.

Boran, Idil, and Corey Katz. 2017. "Climate Change Justice." *Routledge Encyclopedia of Philosophy, Taylor and Francis,* https://www.rep.routledge.com/articles/thematic/climate-change-justice/v-1.

Boran, Idil, and Kenneth Shockley. 2015. "COP 20 Lima: The Ethical Dimension of Climate Negotiations on the Way to Paris–Issues, Challenges, Prospects." *Ethics, Policy & Environment* 18(2): 117–22.

Bulkeley, Harriet, Liliana Andonova, Michele Betsill, Daniel Compagnon, Thomas Hale, Matthew Hoffmann, Peter Newell, Matthew Paterson, Charles Roger, and Stacy VanDeveer. 2014. *Transnational Climate Change Governance*. New York: Cambridge University Press.

Chan, Sander, and Wanja Amling. 2019. "Does Orchestration in the Global Climate Action Agenda Effectively Prioritize and Mobilize Transnational Climate Adaptation Action?" *International Environmental Agreements: Politics, Law and Economics* 19(4–5): 429–46.

Chan, Sander, Idil Boran, Harro van Asselt, Gabriela Iacobuta, Navam Niles, Katharine Rietig, Michelle Scobie, et al. 2019. "Promises and Risks of Nonstate Action in Climate and Sustainability Governance." *Wiley Interdisciplinary Reviews: Climate Change*, 10(e572): 1–8.

Chan, Sander, Clara Brandi, and Steffen Bauer. 2016. "Aligning Transnational Climate Action with International Climate Governance: The Road from Paris." *Review of European, Comparative & International Environmental Law* 25(2): 238–47.

Chan, Sander, Robert Falkner, Matthew Goldberg, and Harro van Asselt. 2018. "Effective and Geographically Balanced? An Output-Based Assessment of Non-State Climate Actions." *Climate Policy* 18(1): 24–35.

Chan, Sander, Harro van Asselt, Thomas Hale, Kenneth W. Abbott, Marianne Beisheim, Matthew Hoffmann, Brendan Guy, et al. 2015. "Reinvigorating International Climate Policy: A Comprehensive Framework for Effective Nonstate Action." *Global Policy* 6(4): 466–73.

Crocker, David A., and Ingrid Robeyns. 2009. "Capability and Agency." In *Amartya Sen*, edited by Christopher W. Morris, 60–90. Contemporary Philosophy in Focus. Cambridge: Cambridge University Press.

Díaz, Sandra, Sebsebe Demissew, Julia Carabias, Carlos Joly, Mark Lonsdale, Neville Ash, Anne Larigauderie, et al. 2015. "The IPBES Conceptual

Framework—Connecting Nature and People." *Current Opinion in Environmental Sustainability* 14(June): 1–16.
Dinerstein, E., A. R. Joshi, C. Vynne, A. T. L. Lee, F. Pharand-Deschênes, M. França, S. Fernando, et al. 2020. "A 'Global Safety Net' to Reverse Biodiversity Loss and Stabilize Earth's Climate." *Science Advances* 6(36): eabb2824.
Dryzek, John S. 2017. "The Meanings of Life for Non-State Actors in Climate Politics." *Environmental Politics* 26(4): 789–99.
Falkner, Robert. 2016. "The Paris Agreement and the New Logic of International Climate Politics." *International Affairs* 92(5): 1107–25.
Ferguson, Yale H., and Richard W. Mansbach. 2004. *Remapping Global Politics: History's Revenge and Future Shock*. Cambridge: Cambridge University Press.
Gardiner, Stephen Mark, and David A. Weisbach. 2016. *Debating Climate Ethics*. New York: Oxford University Press.
Gheaus, Anca. 2013. "The Feasibility Constraint on The Concept of Justice." *The Philosophical Quarterly* 63(252): 445–64.
Gilabert, Pablo. 2012. "Comparative Assessments of Justice, Political Feasibility, and Ideal Theory." *Ethical Theory and Moral Practice* 15(1): 39–56.
Gilabert, Pablo. 2017. "Justice and Feasibility." In *Political Utopias*, edited by Michael Weber and Kevin Vallier, 95–126. Oxford University Press.
Gilabert, Pablo, and Holly Lawford-Smith. 2012. "Political Feasibility: A Conceptual Exploration." *Political Studies* 60(4): 809–25.
Griggs, David, Mark Stafford-Smith, Owen Gaffney, Johan Rockström, Marcus C. Öhman, Priya Shyamsundar, Will Steffen, Gisbert Glaser, Norichika Kanie, and Ian Noble. 2013. "Sustainable Development Goals for People and Planet." *Nature* 495(7441): 305–7.
Gutmann, Amy, and Dennis Thompson. 2004. *Why Deliberative Democracy?* Princeton, NJ: Princeton University Press.
Hale, Thomas. 2016. "'All Hands on Deck': The Paris Agreement and Nonstate Climate Action." *Global Environmental Politics* 16(3): 12–22.
Hale, Thomas, and David Held, eds. 2011. *Handbook of Transnational Governance: Institutions and Innovations*. Cambridge: Polity.
Held, David, and Pietro Maffettone. 2016. "Introduction : Globalization, Global Politics and the Cosmopolitan Plateau." In *Global Political Theory*, edited by David Held and Pietro Maffettone, 1–21. Malden: Polity.
Hickmann, Thomas. 2017. "The Reconfiguration of Authority in Global Climate Governance." *International Studies Review*, July.
Higgott, Richard A., Geoffrey R. D. Underhill, and Andreas Bieler, eds. 2000. *Non-State Actors and Authority in the Global System*. Routledge/Warwick Studies in Globalisation 1. London: Routledge.
Hoffmann, Matthew J. 2011. *Climate Governance at the Crossroads: Experimenting with a Global Response after Kyoto*. Oxford; New York: Oxford University Press.
Höhne, Niklas, Takeshi Kuramochi, Carsten Warnecke, Frauke Röser, Hanna Fekete, Markus Hagemann, Thomas Day, et al. 2017. "The Paris Agreement: Resolving the Inconsistency between Global Goals and National Contributions." *Climate Policy* 17(1): 16–32.

Hsu, Angel, John Brandt, Oscar Widerberg, Sander Chan, and Amy Weinfurter. 2020. "Exploring Links between National Climate Strategies and Non-State and Subnational Climate Action in Nationally Determined Contributions (NDCs)." *Climate Policy* 20(4): 443–57.

Hsu, Angel, Yaping Cheng, Amy Weinfurter, Kaiyang Xu, and Cameron Yick. 2016. "Track Climate Pledges of Cities and Companies." *Nature* 532(7599): 303–6.

Hsu, Angel, Niklas Höhne, Takeshi Kuramochi, Mark Roelfsema, Amy Weinfurter, Yihao Xie, Katharina Lütkehermöller, et al. 2019. "A Research Roadmap for Quantifying Non-State and Subnational Climate Mitigation Action." *Nature Climate Change* 9(1): 11–17.

Jacquemont, Frédéric, and Alejandro Caparrós. 2002. "The Convention on Biological Diversity and the Climate Change Convention 10 Years After Rio: Towards a Synergy of the Two Regimes?" *Review of European Community & International Environmental Law* 11(2): 169–80.

Jamieson, Dale. 2014. *Reason in a Dark Time: Why the Struggle against Climate Change Failed--and What It Means for Our Future.* Oxford; UK: Oxford University Press.

Kanie, Norichika, and Frank Biermann, eds. 2017. *Governing through Goals: Sustainable Development Goals as Governance Innovation.* Earth System Governance. Cambridge: MIT Press.

Keohane, Robert. 2016. "Keohane on Climate: What Price Equity and Justice?" *Climate Home News* (blog). September 6, 2016. Retrieved from: http://www.climatechangenews.com/2016/09/06/keohane-on-climate-what-price-equity-and-justice/.

Klinsky, Sonja, Timmons Roberts, Saleemul Huq, Chukwumerije Okereke, Peter Newell, Peter Dauvergne, Karen O'Brien, et al. 2017. "Why Equity Is Fundamental in Climate Change Policy Research." *Global Environmental Change* 44(May): 170–73.

Krasner, Stephen D., ed. 1983. *International Regimes.* Cornell Studies in Political Economy. Ithaca: Cornell University Press.

Kroll, Christian, Anne Warchold, and Prajal Pradhan. 2019. "Sustainable Development Goals (SDGs): Are We Successful in Turning Trade-Offs into Synergies?" *Palgrave Communications* 5(1): 140.

Langenbrunner, Baird. 2020. "Hazards in the Himalayas." *Nature Climate Change* 10(5): 385.

Lawford-Smith, Holly. 2013. "Understanding Political Feasibility" *Journal of Political Philosophy* 21(3): 243–59.

Lomborg, Bjørn, ed. 2010. *Smart Solutions to Climate Change: Comparing Costs and Benefits.* Cambridge: Cambridge University Press.

Maes, Frank, A. Cliquet, Willemien Du Plessis, and Heather McLeod-Kilmurray, eds. 2013. *Biodiversity and Climate Change: Linkages at International, National and Local Levels.* The IUCN Academy of Environmental Law Series. Cheltenham, UK: Edward Elgar.

Mal, Suraj, R. B. Singh, Christian Huggel, and Aakriti Grover. 2018. "Introducing Linkages Between Climate Change, Extreme Events, and Disaster Risk Reduction."

In *Climate Change, Extreme Events and Disaster Risk Reduction*, edited by Suraj Mal, R.B. Singh, and Christian Huggel, 1–14. Sustainable Development Goals Series. Cham: Springer International Publishing.

Maljean-Dubois, Sandrine. 2016. "The Paris Agreement: A New Step in the Gradual Evolution of Differential Treatment in the Climate Regime?" *Review of European, Comparative & International Environmental Law* 25(2): 151–60.

Marselle, Melissa R, Jutta Stadler, Horst Korn, Katherine N Irvine, and Aletta Bonn. 2019. *Biodiversity and Health in the Face of Climate Change.* https://www.springer.com/gp/book/9783030023171.

Moellendorf, Darrel. 2014. *The Moral Challenge of Dangerous Climate Change: Values, Poverty, and Policy.* New York: Cambridge University Press.

Morita, Kanako, and Ken'ichi Matsumoto. 2015. "Enhancing Biodiversity Co-Benefits of Adaptation to Climate Change." In *Handbook of Climate Change Adaptation*, edited by Walter Leal Filho, 953–72. Berlin, Heidelberg: Springer Berlin Heidelberg.

Ngwadla, Xolisa. 2014. "An Operational Framework for Equity in the 2015 Agreement." *Climate Policy* 14(1): 8–16.

Nilsson, Måns, Elinor Chisholm, David Griggs, Philippa Howden-Chapman, David McCollum, Peter Messerli, Barbara Neumann, Anne-Sophie Stevance, Martin Visbeck, and Mark Stafford-Smith. 2018. "Mapping Interactions between the Sustainable Development Goals: Lessons Learned and Ways Forward." *Sustainability Science* 13(6): 1489–503.

Nussbaum, Martha C. 2011. *Creating Capabilities: The Human Development Approach.* Cambridge, Mass: Belknap Press of Harvard University Press.

Okereke, Chukwumerije, Harriet Bulkeley, and Heike Schroeder. 2009. "Conceptualizing Climate Governance Beyond the International Regime." *Global Environmental Politics* 9(1): 58–78.

Pattberg, Philipp. 2010. "Public-Private Partnerships in Global Climate Governance: Public-Private Partnerships." *Wiley Interdisciplinary Reviews: Climate Change* 1(2): 279–87.

Patterson, James J, Thomas Thaler, Matthew Hoffmann, Sara Hughes, Angela Oels, Eric Chu, Aysem Mert, Dave Huitema, Sarah Burch, and Andy Jordan. 2018. "Political Feasibility of 1.5°C Societal Transformations: The Role of Social Justice." *Current Opinion in Environmental Sustainability* 31(April): 1–9.

Posner, Eric A., and David A. Weisbach. 2010. *Climate Change Justice.* Princeton: Princeton University Press.

Rossati, Antonella. 2017. "Global Warming and Its Health Impact." *The International Journal of Occupational and Environmental Medicine* 8(1): 7–20.

Ruggie, John Gerard. 2004. "Reconstituting the Global Public Domain—Issues, Actors, and Practices." *European Journal of International Relations* 10(4): 499–531.

Sen, Amartya. 1999. *Development as Freedom.* New York: Anchor Books.

Shockley, Kenneth. 2014. "Sourcing Stability in a Time of Climate Change." *Environmental Values* 23: 199–217.

Shockley, Kenneth. 2018. "The Great Decoupling: Why Minimizing Humanity's Dependence on the Environment May Not Be Cause for Celebration." *J Agric Environ Ethics* 31: 429–442.

Shue, Henry. 1999. "Global Environment and International Inequality." *International Affairs* 75(3): 531–45.
Shue, Henry. 2010. "Deadly Delays, Saving Opportunities: Creating a More Dangerous World?" in *Climate Ethics*, ed. Stephen M. Gardiner, Simon Caney, Dale Jamieson, and Henry Shue, 146–62. New York: Oxford.
Southwood, Nicholas. 2018. "The Feasibility Issue." *Philosophy Compass* 13(8): e12509.
Turral, Hugh, Jacob J. Burke, and Jean-Marc Faurès. 2011. *Climate Change, Water and Food Security*. FAO Water Reports 36. Rome: Food and Agriculture Organization of the United Nations.
UN (United Nations). 2015. Transforming our World: The 2030 Agenda for Sustainable Development. Retrieved from: https://stg-wedocs.unep.org/bitstream/handle/20.500.11822/11125/unep_swio_sm1_inf7_sdg.pdf?sequence=1
UNFCCC (United Nations Framework Convention on Climate Change). 2015. Adoption of the Paris Agreement, Decision 1/CP. 21 FCCC/CP/2015/10/Add.1 Retrieved from: https://unfccc.int/sites/default/files/resource/docs/2015/cop21/eng/10a01.pdf
Van Asselt, Harro. 2014. *The Fragmentation of Global Climate Governance: Consequences and Management of Regime Interactions*. New Horizons in Environmental and Energy Law. Cheltenham: Edward Elgar.
Van Asselt, Harro. 2016. "Editorial." *Review of European, Comparative & International Environmental Law* 25(2): 139–41.
Wapner, Paul K. 1996. *Environmental Activism and World Civic Politics*. SUNY Series in International Environmental Policy and Theory. Albany: State University of New York Press.
Watts, Nick, W Neil Adger, Paolo Agnolucci, Jason Blackstock, Peter Byass, Wenjia Cai, Sarah Chaytor, et al. 2015. "Health and Climate Change: Policy Responses to Protect Public Health." *The Lancet* 386(10006): 1861–914.
Young, Oran R. 1989. *International Cooperation: Building Regimes for Natural Resources and the Environment*. Cornell Studies in Political Economy. Ithaca: Cornell University Press.
Young, Oran R. 2017. *Governing Complex Systems: Social Capital for the Anthropocene*. Earth System Governance. Cambridge: The MIT Press.

Chapter 3

Climate Justice in the Nonideal Circumstances of International Negotiations

Michel Bourban

Since climate justice emerged as a research field in the early 1990s, scholars have focused a great deal on providing a rationale for principles of justice. Burden-sharing justice (based on the polluter pays, the ability to pay, and the beneficiary pays principles) and harm avoidance justice (based on the harm principle) represent the two major kinds of climate justice.[1] Scholars have especially used idealizations, theoretical models, and thought experiments in an attempt to find the fairest way to share the global and intergenerational burdens of climate change, in light of historical responsibility for past emissions.[2]

But how can these abstract philosophical principles be translated into concrete climate policies? Although climate policies were already considered in early debates on climate justice, over the last decade, growing attention has been given to fair and feasible institutional reforms. There are two important reasons for this recent development of action-guiding approaches to climate justice. The first is that, since the 1990s, the gap between the demands of climate justice and existing climate policies has widened. Despite further theoretical justification for the principles of climate justice, political inertia still reigns at the national and international levels. As Steve Vanderheiden (2016, 28) laments:

> The gap between what scholars have called for as a matter of climate justice and what is politically and institutionally feasible has grown, with ideal theory work on environmental justice ironically making its own prescriptions appear to be decreasingly obtainable in consequence of their widening distance from the practical political means available for bringing them about.

The second reason for the growing chasm between theories of climate justice and climate policy is the rising awareness that we are in a situation of climate emergency in which ambitious political action is urgently needed. The IPCC (2018, 12) warns us that for a 66 percent probability of limiting global warming to 1.5°C, humanity has a remaining carbon budget of no more than 420 GtCO_2. At current emission rates (42 GtCO_2 per year), this budget is expected to be exhausted in about ten years.[3] The growing gap between (ideal) theory and practice, together with the rapidly shrinking global carbon budget, has pushed normative political theorists to become increasingly interested in finding more efficient and fairer climate change mitigation, adaptation, and compensation policies (Maltais 2016, Light and Taraska 2016, Heyward and Ödalen 2016, Page and Heyward 2016, Vanderheiden 2016, Baatz 2018, Bourban 2018).

I contribute to these recent developments in the philosophical literature on climate change by showing one possible way to promote climate justice in the nonideal circumstances of international negotiations. In the first section, I develop methodological considerations regarding ideal and nonideal theory and define the notion of political feasibility. In the second section, I present an institutional reform to the international climate regime focused on mitigation responsibilities. I argue that an equity calculator based on the norm of common but differentiated responsibilities and respective capabilities (CBDR-RC) should be utilized by the climate regime, specifically in countries' nationally determined contributions (NDCs). In the third section, I explain how this proposal is feasible by focusing on three main arguments: the flexibility of the proposed normative framework, the role of civil society, and the importance of states' reputational concerns in the bottom-up architecture of the Paris Agreement.

My objective is to follow the methodological change of course recently initiated by some political philosophers in order to develop a nonideal approach to climate justice that addresses the feasibility issue. The high degree of compatibility between my reform proposal and existing political institutions within the climate regime, especially the fundamental norm of CBDR-RC and the five-year cycle for the reevaluation of NDCs set by the Paris Agreement, ensures a high degree of feasibility. At the same time, the congruence between the equity calculator and basic principles of climate justice promotes a higher degree of fairness. Moral and political theorizing is not constrained by feasibility constraints, which is why ideal theory matters; however, since the feasible set constraints the space within which we can apply principles of justice, it is important to complement existing ideal theories of climate justice with nonideal approaches to climate justice.

I. TOWARD A NONIDEAL APPROACH TO CLIMATE JUSTICE

Ideal and Nonideal Theory

Although there are many reasons for the growing gap between the demands of justice and existing climate policies, one important factor is the dominant position of ideal theory in the field of climate justice. Originally, it made sense to follow Rawls's methodology and propose different and competing grand theories of climate justice (liberal, cosmopolitan, utilitarian, etc.). However, now we can see that the normative prescriptions of the first wave of debates on climate justice have had little influence on real-world politics, and so a methodological change of course seems warranted. As Laura Valentini (2012, 654) points out, "[s]ome [political philosophers] voiced the concern that the dominant—Rawlsian—paradigm in the discipline was somehow too detached from reality to guide political action." She continues: "[t]his methodological debate on the proper nature of political philosophy, and its ability to guide action in real-world circumstances, has become known as the debate on ideal and non-ideal theory" (ibid.). This chapter is a contribution to a nonideal approach to climate justice that aims at making the normative prescriptions of the field more action-guiding.[4] The objective is not to question the relevance of ideal theories of climate justice, on which I draw in part, but rather to complement them. In this section, I explain in what sense a nonideal approach is better suited to dealing with the topic of feasibility.

There are several ways to understand the much-debated methodological distinction between ideal and nonideal theory.[5] Originally, Rawls (1999, 216) conceived ideal theory as the part of a theory of justice assuming "strict compliance" for the purpose of working out the principles "that characterize a well-ordered society under favorable circumstances." He stated that his "main concern" was "with this part of the theory" (ibid.).

By following Rawls's conception of ideal theory, theorists of (climate) justice often make two major idealizations (Chung 2012, 65):

(1) Compliance with principles of justice: citizens and political leaders are reasonable beings using their sense of justice and their practical reason to comply with the demands of justice.
(2) The existence of circumstances favorable to the achievement of justice: principles of justice are readily applicable to existing institutions.

The claim is not that people are *in fact* reasonable beings or that the current situation is *in fact* favorable to the realization of justice, but those are *assumed* for the purpose of finding an ideal theory. Rawls (1999, 8, emphasis added) explains that "everyone is *presumed* to act justly and to do his part in upholding just institutions." Such assumptions are valuable in the exploration of "what a perfectly just society would look like" (Rawls 1999, 8).

Rawls (1999, 216) conceived ideal and nonideal theory as two complementary subparts of the same theory of justice. Once an ideal theory has been formulated, nonideal theory can start to explore how to deal with actual injustices. The nonideal part of the theory moves from "the conception of a perfectly just basic structure" to the real world, and addresses problems "under less happy conditions," such as partial compliance and noncompliance (Rawls 1999, 216–217).

My aim is not to fully investigate the methodological distinction between ideal and nonideal theory, but rather to draw on nonideal theory to think about climate justice. While there are still important disagreements between climate justice scholars on the best set of principles of distributive justice (Caney 2010, Page 2011), I presuppose that some basic principles of climate justice as defined by ideal theory, such as the polluter pays principle (PPP), the ability to pay principle (APP), and the beneficiary pays principle (BPP), are ethically relevant. As Henry Shue (2015, 8) explains, "Even though some may diverge at the theoretical periphery," key principles of climate justice "converge at the practical core." There is indeed a broad ethical consensus that developed countries should shoulder most mitigation burdens, at least initially: "There is enough ethical convergence to provide guidelines for 'rough justice' moving forward, at least in the initial steps" (Gardiner and Weisbach 2016, 129). While it is of course possible to further justify basic principles of justice in ideal theory, the priority today is for developed countries to implement ambitious mitigation policies and to help developing and emerging countries to mitigate their own emissions.[6]

In other words, nonideal justice directs us, at least in the initial steps of the transition to fairer climate policies, in the same general direction as principle of climate justice do: affluent countries should bear most of the burden and play a leadership role. In a second step, a nonideal approach can increase the level of specificity and show that the PPP and the APP can lead to different results, for instance, in the case of high emitting but poor counties. However, both with the first and the second steps, a nonideal approach to climate justice follows the demands of principles of burden-sharing and harm avoidance justice.

Two major sets of problems are characteristic of a nonideal approach. First, partial compliance and noncompliance with principles of justice. These can be caused by the limited and fallible moral reasoning of moral agents. They can also be motivated by a plurality of other reasons, including amoral motives, such as the promotion of personal interests. The second set of problems is due to the existence of circumstances that are unfavorable to the realization of just institutions. The case of climate change illustrates this well. A first problem here is institutional inadequacy, caused by the "limits of our current, largely national, institutions, and the lack of an effective system of global governance" (Gardiner 2011, 29). A second problem is the existence of unjust institutions and practices, such as global subsidies for fossil fuels, which reached $4.9 trillion in 2013, and $5.3 trillion in 2015—a total of 6.5 percent of global GDP for both years (Coady et al. 2017).

As I understand it here, a nonideal approach to climate justice involves two major tasks related to these two sets of problems. The first task is to address the problems of partial compliance and noncompliance with principles of justice. What should be done when individual and collective agents do not follow the demands of climate justice? One way to approach this is to investigate quasi-moral and amoral reasons to fight against climate change, stressing the convergence between moral, quasi-moral, and amoral motives in addressing the climate problem (Bourban 2014, André and Bourban 2016). In this chapter, however, I focus on the second task of nonideal theory: proposing institutional reforms that are both fair and feasible.[7] A nonideal approach to climate justice aims to be action-guiding or practice-oriented, especially policy-oriented. Its three main characteristics are

(1) a high degree of fact-sensitivity,
(2) an interest in transitional improvements, and
(3) an incorporation of relevant feasibility constraints.[8]

These three characteristics can be understood as the key components of what Stephen Gardiner (2011, 400) more broadly calls "the ethics of the transition," whose objective is to "articulate how we might proceed ethically starting from existing, and sometimes deeply constrained or ethically compromised, social realities in the direction of better solutions and general circumstances." They also reflect what Henry Shue (1996, 16–17) named "extrication ethics," or "transition ethics." Although "sometimes the best route to the ideal is to set an uncompromising example of the ideal," more often, we cannot "simply assume that good examples will routinely inspire emulation, whatever the circumstances" (ibid.). This is why, in addition to offering "an ultimate ideal" and merely wishing individual and collective agents "good luck in figuring how to reach it," political theorists should also offer "principles for transition" that are "directly applicable to how change is brought about" (ibid.).

Defining Feasibility

Political philosophy that aspires to practical relevance must engage with feasibility constraints. But what is "feasibility?" To begin with, there are two main types of feasibility.⁹ On the one hand, some feasibility constraints are "hard," for instance logical, physical, and biological laws. Climate sciences present such hard constraints. Take the example of the carbon budget. For any limit in global temperature increase, there is a corresponding amount of greenhouse gases (GHGs) that can be emitted into the atmosphere. The carbon budget for keeping global temperature increase below 1.5°C or 2°C is not negotiable.¹⁰ As James Hansen (2009, 206) stresses, "Nature and the laws of physics cannot compromise—they are what they are." On the other hand, some parameters of feasibility are rather "soft," in the sense that they are malleable and hold only for certain agents and time spans. Soft feasibility constraints are economic, political, and cultural. They are typically addressed by researchers in social and political sciences, when they describe the rules, practices, procedures, and conventions that govern international relations.

A nonideal approach to climate justice has to address both types of constraints. It must draw on data on the physical underpinnings of the climate system, and on how climate negotiations work, how the institutions involved function. Accurate practical and ethical guidance cannot be offered without sound knowledge of both the physical dynamics of climate change and the political dynamics of international negotiations. The two kinds of empirical information are also related. Data on hard feasibility constraints such as the scarce and rapidly decreasing global carbon budget should spur normative political theorists to take existing political constraints seriously. The chances of avoiding dangerous anthropogenic interference with the climate system are rapidly dwindling. Hence, any possibility of implementation in the near future means practical recommendations need to start from where we are now, rather than where we want to be.

In contrast with hard feasibility constraints, soft feasibility constraints can evolve over time, for instance, through technological innovations, education, or cultural transformations. Political, economic, and cultural constraints come in different degrees. Judgments of feasibility are indexed to circumstances and time: demands that are infeasible in certain circumstances or at a certain time may prove to be feasible in different circumstances or times. The feasible set can actually evolve quite rapidly. As Gardiner points out, "[i]n my lifetime, many things previously touted as infeasible by seasoned political commentators and conventional wisdom have occurred (e.g., the fall of the Berlin Wall, the peaceful collapse of apartheid, the election of a black President)" (Gardiner and Weisbach 2016, 54). We could add to this list the recent political measures taken to face the coronavirus pandemic, many of which would have been considered infeasible before the start of the crisis. At the time of writing, multiple regions throughout the world are still under lockdown or subject to strict social distancing measures, EU leaders have agreed on a common emergency plan worth €750 billion (Boffey and Rankin 2020), and in the United States the stimulus package is expected to reach around $2 trillion (McKelvey 2020). (See Araujo and Meyer, L. in volume 1 for more on this topic.) All these developments are contributing to radically change the political, economic, and social landscape.[11]

Considerations of political feasibility can all too easily be used as an excuse for postponing actions that would allow injustices to be reduced. Because they are perceived as infeasible, ambitious actions to mitigate global GHG emissions may be neglected. Judgments of (in)feasibility can reflect a lack of political willingness, rather than a lack of ability, to act more quickly and face climate injustices more radically. This is why it is important not to mischaracterize the feasible set, which delineates the realm of what is possible here and now.

Conceiving the Feasible Set

Painting a distorted picture of the feasible set can lead to pragmatic considerations unnecessarily limiting the pursuit of climate justice. If disproportionate weight is given to some feasibility constraints, just solutions may be excluded from consideration. Eric Posner and David Weisbach's "new realism" (Ohlin 2015, 8) clearly illustrates this problem. On the grounds that philosophers lack understanding of the realities of international relations, they "reject the claim that certain intuitive ideas about justice should play a major role in the design of a climate agreement" (Posner and Weisbach 2010, 5). Such ideas are problematic because "they fail to consider basic pragmatic or feasibility constraints" (Posner and Weisbach 2010, 4). Posner and Weisbach argue that the only feasible climate treaty is one that is in the short-term domestic self-interest of developed countries, particularly the United States. By doing so, they dismiss most climate justice demands as infeasible because they would be too costly for rich countries, among other reasons. (See Weisbach in volume 1 for more on this topic.)

The key feasibility constraint isolated by Posner and Weisbach (2010, 6) is International Paretianism, according to which "treaties are not possible unless they have the consent of all states, and states only enter treaties that serve their interests." Simon Caney (2014, 128) breaks down their argument the following way:

(P1) It is necessary to have a climate treaty with which major emitters comply.
(P2) To be feasible an effective climate treaty must serve the interests of high emitting states (from "Feasibility" to "Pareto Superiority"). Therefore,
(C) A climate treaty must serve the interests of high emitting states.

One major problem raised by this argument is that it only adopts the internal perspective of high-emitting states, and not the external perspective of other states trying to convince major emitters to mitigate their emissions (Caney 2014, 130). With its overly narrow assumption that to be feasible, any climate treaty must necessarily be in the short-term, self-perceived economic interest of developed states, this position develops an inaccurate picture of the feasible set. Consequently, Posner and Weisbach (2010, 86) do not only exclude just and feasible solutions; they also support unjust solutions by claiming that an "optimal climate treaty" could require side payments to rich countries from very poor and vulnerable countries, thereby advocating a pollut*ed* pays principle in lieu of a PPP.[12] This position is not only unfair but also ultimately infeasible, being unacceptable to multiple Parties to the United Nations Framework Convention on Climate Change (UNFCCC), an international regime in which any political and legal decision has to be made by consensus.[13]

This overly restrictive conception of the feasible set comes from a very controversial way of viewing international relations. Jens Ohlin (2015, 9) explains that "Posner is, in many respects, the modern father of the new skepticism about international law." Posner's new realism seeks to systematically undermine the foundations of international law by arguing that states are motivated solely by their short-term economic interests and that not following international legal norms is justified as soon as they conflict with narrowly perceived national interests. By convincing a whole range of legal scholars and policymakers that only international rules, principles, and agreements that benefit the United States and its geopolitical allies in the short term are acceptable, Posner has contributed to spreading an excessively narrow idea of what is feasible and what is not.

It is important therefore to be very careful not to let considerations of feasibility limit the pursuit of climate justice. Feasibility does not impose limits on moral and political theorizing; it merely constrains the space within which we can pursue normative goals (Bell 2013, Roser 2015). The point is not to change how we justify principles of climate justice, or to question their ethical and political relevance, but rather to find ways to apply them, once justified, to the nonideal circumstances of the real world. This means explaining how principles of climate justice can be operationalized (see section 2), and then clarifying in what sense the proposed operationalization is feasible (see section 3).

II. CLOSING THE AMBITION GAP

Ethics in the Climate Regime

One crucial task for a nonideal approach is to find feasible ways to help bridge the gap between existing climate policies and the demands of burden-sharing and harm avoidance justice. I focus here on mitigation policies and propose one possible institutional reform that could contribute to closing the ambition gap in global climate policy.

The climate regime is a complex and evolving structure of norms and institutions that frame international negotiations on climate change. The notion of "regime" refers not only to legal and political structures but also to the modes of production of scientific knowledge (Aykut and Dahan 2015, 67–69). In both the construction and the development of the climate regime, politics and science constantly interact. The climate regime includes the UNFCCC, which was signed in 1992 and entered into force in 1994; the Conference of the Parties (COPs), which began in 1995 in Berlin (COP1) and last took place in Madrid in 2019 (COP25); and the assessment and special reports of the Intergovernmental Panel on Climate Change (IPCC), the first of which was published in 1990 (First Assessment Report) to the recent 2019 report (Special Report on the Ocean and Cryosphere).

The climate regime makes substantial room for ethical considerations. In its Fifth Assessment Report,[14] the Working Group III (2014b, 5) for instance states that

> Issues of equity, justice, and fairness arise with respect to mitigation and adaptation. . . . Many areas of climate policy-making involve value judgements and ethical considerations. These areas range from the question of how much mitigation is needed to prevent dangerous interference with the climate system to choices among specific policies for mitigation or adaptation.

The IPCC Special Report on Global Warming of 1.5°C even puts ethics at the core of its concern:

> Ethical considerations, and the principle of equity in particular, are central to this report, recognizing that many of the impacts of warming up to and beyond 1.5°C, and some potential impacts of mitigation actions required to limit warming to 1.5°C, fall disproportionately on the poor and vulnerable (*high confidence*). Equity has procedural and distributive dimensions and requires fairness in burden sharing both between generations and between and within nations. (Allen et al. 2018, 51)

When one moves from the scientific to the political component of the climate regime, one can observe that normative language has also been used since the start of climate diplomacy. The parties to the UNFCCC (1992, art. 3.1) agreed that their core objective would be to "protect the climate system for the benefit of present and future generations of humankind, on the basis of equity and in accordance with their common but differentiated responsibilities and respective capabilities [CBDR-RC]." This norm of CBDR-RC plays a major role in all the main outcomes of climate negotiations, including in the Paris Agreement. Christian Holz, Sivan Kartha, and Tom Athanasiou (2018, 119) point out that "it is the principle of CBDR/RC that has been the pivot of most equity debates within the climate negotiations."

Despite these discussions on ethical notions such as equity, responsibility, and capacity, the existence of an ambition gap between the demands of climate justice and current climate policies shows that normative considerations have had limited impact on policymaking. In addition to ethical motives, other considerations have shaped climate policy over the past three decades, especially short-term economic thinking by fossil fuel providers and consumers. Should we therefore concede to realists that ethical arguments and normative beliefs are causally irrelevant in international relations?

Constructivist international relations theorists such as Neta Crawford (2002, 2) reply to realists that "ethical argument analysis is a way to understand and explain normative change in world politics." Because they posit new normative beliefs and new behaviors, "ethical arguments may change actors' conceptions of what is possible or what is desirable" (ibid.). These arguments "can even result in actors reframing their 'interests,' or the order of their preferences" (Crawford 2002, 103). It is in the interest of states to comply with norms, such as CBDR-RC, because international law represents, as Ohlin (2015, 54) explains, a "legal system" of "mutually beneficial constraints." Complying with such international norms is a rational behavior, because it provides "meaningful benefits for all involved" (Ohlin 2015, 54), such as avoiding a dangerous anthropogenic interference with the climate system.

It is not an easy thing to change normative beliefs and behavioral norms. However, Crawford (2002, 116) highlights that "[e]thical appeals work well when their content is linked to the dominant belief systems, social institutions, and identities of actors." This is why a fundamental criterion of feasibility for any institutional reform proposal is to achieve a sufficiently high degree of compatibility with existing political norms and institutions. Since CBDR-RC represents the cornerstone of the climate regime, it should be taken as a starting point and properly operationalized. CBDR-RC can converge with the interests of powerful states insofar as they hold that it is rational for them to act so as to avoid dangerous climate change. Even if they interpret this international norm in a self-interested way, ethical arguments can change their conceptions of what is possible or desirable.

A legally binding procedural tool has been included in the Paris Agreement to push countries to meet the demands of CBDR-RC when they define their national commitments. Every five years, each country has to submit a new, more ambitious NDC, in accordance with CBDR-RC[15]:

Each Party's successive nationally determined contribution will represent a progression beyond the Party's then current nationally determined contribution and reflect its highest possible ambition, reflecting its common but differentiated responsibilities and respective capabilities, in the light of different national circumstances. (UNFCCC 2015, art. 4.3)

If it is properly implemented, this five-year cycle could represent an opportunity to reduce the ambition gap in climate policies. A major challenge today is to redefine "the rules and principles that will make it possible to coordinate these national actions, to avoid them being too disparate and too unambitious" (Aubertin et al. 2015, 4). As a contribution to meet this challenge, my proposal is to introduce into the climate regime a normative framework for assessing the degree of equity of NDCs and comparing the degree of ambition of national mitigation pledges between them. The point is not to propose a specific definition of the principles of climate justice. Rather, it is to develop a flexible framework with which NDCs must comply if they are to meet the demands of CBDR-RC. To be more than a confrontation between various divergent and irreconcilable national positions, the negotiation process must follow a transparent model based on the fundamental principles that states have committed to respecting. A climate justice index could represent a key element of such a model.

Operationalizing CBDR-RC

There are several proposals for translating CBDR-RC into a burden-sharing formula. In the following, I review only one such proposal that serves the purpose of a nonideal approach to climate justice, and that responds to realist-minded objections based on considerations of feasibility.[16] The Greenhouse Development Rights framework (GDR) is based on an index composed of a responsibility indicator (cumulative national emissions since 1990) and a capacity indicator (per capita annual income above a development threshold of $8,500) (Baer et al. 2009, Baer 2013). This responsibility-capacity index determines the percentage of total global obligation for each country by giving each indicator the same weight. As a result, in 2010, the United States held 29.4 percent of global obligation, the EU 26 percent, Japan 7.6 percent, Russia 5.8 percent, China 5.1 percent, Brazil 2.8 percent, and India and South Africa 0.9 percent each. In total, high-income countries held 73.7 percent of global obligation, least developed countries (LDCs) 0.3 percent, and the remaining 26 percent was held by new emitting countries, such as countries from the Brazil, South Africa, India, and China (BASIC) group (Baer 2013).

Recognizing that their choice of indicators is debatable according to different countries' interpretation of CBDR-RC, the GDR framework has recently been updated to create the "Climate equity reference framework" (CERF), which proposes a "Climate equity reference calculator" to assign each country and region its fair share of mitigation efforts (CERP 2015, Holz, Kartha, and Athanasiou 2018). In contrast to the GDR, the CERF includes an equity band that allows more or less-demanding effort-sharing parameters to be chosen, giving countries more flexibility to calculate their respective level of responsibility and capacity. Three key parameters are flexible. First, the level of global ambition is flexible, with three possible mitigation pathways: 1.5°C Low Energy Demand (66 percent chance of 1.5°C in 2100); 1.5°C Standard (greater than or equal to 50 percent chance of staying below 1.5°C in 2100); 2°C Standard (greater than 66 percent chance of staying within 2°C in 2100).[17] Second, the historical responsibility parameter is flexible, with possible start dates ranging from 1850 to 2010. Third, the capacity parameter is flexible, with the option to set the development threshold between $0 and $20,000.

For a 1.5°C-compliant global mitigation effort (1.5°C Standard mitigation pathway), a responsibility indicator with a start date ranging from 1850 to 1990 (the equity band for the responsibility parameter), and a development threshold ranging from $2,500 to $7,500 (the equity band for the capacity parameter), the CERF authors made the following findings.[18] Among high-capacity and high-responsibility countries, such as the United States, European Union, and Japan, "NDCs fall far short of the fair-share contributions as bounded by the equity band" (Holz, Kartha, and Athanasiou 2018, 127). This means, whatever the choice of parameters, these countries fail to assume their global responsibilities. The United States pledged only 16–24 percent of its fair-share contributions, the EU 21–23 percent, and Japan about 10 percent (ibid.). In addition, fair shares of the global effort to mitigate emissions are higher than plausible domestic reductions in these countries (Holz, Kartha, and Athanasiou 2018, 128). This means that, to fulfill their mitigation efforts, they must contribute, through climate finance, to mitigation opportunities in other countries where mitigation potential exceeds domestic obligations. Developed countries therefore have a "dual obligation." In addition to domestic emissions reductions, they are also obligated to engage in international mitigation cooperation.

If we take the norm of CBDR-RC seriously, most developed countries' pledges are so far largely insufficient and, therefore, unfair. This remains true even if we choose undemanding indicators of responsibility and capacity. Integrating the CERF equity calculator in countries' future NDCs could contribute to closing this ambition gap. Other climate justice indices could also be used to that end. Even if disagreements remain on the kinds of indicators that should be used to measure responsibility and capacity, there is a broad agreement on the fact that developed countries are the most responsible and capable, that the current NDCs of these countries are largely unfair, and that they have a dual obligation to mitigate their emissions and help other countries to do so.[19]

Two clarifications are in order here. First, where should this equity calculator be "integrated"? And how could it be concretely used by parties to the UNFCCC? From a global perspective, the equity calculator could be integrated into the climate regime. As stressed above, politics and science have been in constant interaction since climate negotiations began. Scientific experts could play an important role by adding an *Equity Gap Report* to the *Emissions Gap Report* (UNEP 2019) and the *Adaptation Gap Report* (UNEP 2018). The *Emissions Gap Report* measures how countries' actions and pledges affect global GHG emissions trends compared to emissions trajectories consistent with the long-term goal of the UNFCCC. The *Adaptation Gap Report* measures the gap between current adaptation policies and adaptation needs, especially in vulnerable countries. As a complement to these two scientific reports, an *Equity Gap Report* would measure the gap between NDCs and the goal set by the Paris Agreement, based on the CERF equity calculator. It would list and rank countries according to their degree of compliance with the norm of CBDR-RC, with different scores for each country according to different equity benchmarks set by an equity band. Such considerations could also be added by way of a new chapter in the *Emissions Gap Report*, in much the same way as the IPCC included a whole chapter on normative issues in its last Assessment Report (Working Group III 2014a, 283–350).

At the national level, the parties to the UNFCCC could directly integrate the CERF normative framework into their NDCs. The equity calculator is freely available online, and the different countries could use it to explain how their pledges are consistent with the demands of CBDR-RC. If their choice of responsibility and capacity indicators is unreasonable (for instance, setting the development threshold too low or the start date for national emissions too late), this would be reflected in their ranking in the *Equity Gap Report*. Including considerations on the equity calculator in their NDCs would give parties a common and transparent framework to explain in what sense they perceive their mitigation pledges to be fair. It would also encourage countries whose NDCs fall far short of their fair-share contributions to choose more ambitious mitigation goals. The IPCC Working Group III (2014b, 5) states that "[i]ssues of equity, justice, and fairness arise with respect to mitigation and adaptation," and that "[m]any areas of climate policy-making involve value judgements and ethical considerations." The equity calculator would help parties to the UNFCCC to make the value judgments and ethical considerations guiding their mitigation pledges explicit, and to justify them on the basis of CBDR-RC, as the Paris Agreement requires (UNFCCC 2015, art. 4.3).

A second question raised by the proposal is the following: does CBDR-RC align perfectly with principles of climate justice? Is CBDR-RC all that justice demands? In early debates on climate justice, Caney (2005, 773–774) was already voicing some concerns on this matter. On the one hand, CBDR-RC follows the PPP and the APP by insisting that duties to mitigate fall on all, but more on some than others, by ascribing responsibilities according to what the parties have done and what they are able to do. On the other hand, CBDR-RC "tends to be interpreted in such a way that states are held accountable for the decisions of earlier generations" (Caney 2005, 774), without addressing possible objections against historical responsibility for past emissions, such as the argument of excusable ignorance. (For more on the topic of historical responsibility, please see Meyer, L. in this volume.)

Andrew Light (2017, 491) addresses this kind of concern by stressing that CBRD-RC represents a "basic formula for distributive justice," based on each party's current and historical emissions profile and on its relative development needs. Although this norm remains vague, "it is appealed to as providing some of the most explicit guidance for how burdens for mitigation should be distributed between developed and developing countries" (Light 2017, 488). Its vagueness allows room for different interpretations of responsibility and capacity that meet the demands of equity. There is therefore no perfect alignment between CBDR-RC and principles of distributive climate justice, but as explained above, we find ourselves in a situation in which nonideal justice is sufficient to guide mitigation policies. Ideal climate justice would demand more than the CERF understanding of CBDR-RC, but the institutional reform proposed here would make it possible to get closer to climate justice. In other words, the CERF does not represent an account of perfect justice, but in the spirit of a nonideal approach, this normative framework is characterized by a high degree of fact-sensitiveness and provides a strong basis for transitional improvements that would contribute to reducing climate injustices. But in what sense can we say that the proposal to include a climate justice index similar to the one developed by the CERF into the climate regime and/or NDCs is feasible?

III. POLITICALLY FEASIBLE?

In this section, I will explain why the institutional reform proposed in the previous section can be considered feasible. My three main arguments are the following:

(1) The equity calculator is sufficiently flexible to make room for different interpretations of CBDR-RC, but at the same time the equity band ensures that it remains robust enough;

(2) Although this kind of institutional reform is very likely to meet some form of resistance, we can count on it being promoted by counterpowers to fossil energy lobbies and climate laggards;
(3) In the bottom-up architecture of the Paris Agreement, reputational concerns will likely drive many countries to include an equity calculator into their NDCs once other countries have started to do so.

A Flexible, but Robust Normative Framework

A first reason why the inclusion of an equity calculator in countries' NDCs is a feasible reform is that it is based on the cornerstone of the climate regime, the norm of CBDR-RC. As explained above, a fundamental criterion of feasibility is to achieve a sufficiently high degree of compatibility with existing political institutions. But the vagueness of this norm is both its strength and its weakness. The parties to the UNFCCC nominally agreed to CBDR-RC but (a) it can be interpreted in multiple ways and (b) states can interpret it in ways that suit their self-interest. For instance, some countries might emphasize the "common" aspect and reject interpretations of the "differentiated" that place any weight on past emissions and historic responsibility. Thus, a nonideal approach to climate justice faces the following general challenge: either the transitional principle is too lax and major actors tend to comply with it, but it is too vague to guide ambitious action; or the transitional principle is too demanding, in which case it can lead to ambitious action, but it becomes less feasible and major actors are less likely to comply with it.[20]

Finding a middle ground between flexibility and robustness can help to overcome this dilemma. The CERF equity calculator is a transitional principle designed to allow countries to choose more or less-demanding effort-sharing parameters so that they have enough flexibility to calculate their level of global obligation. Since the Paris Agreement often stresses the "different national circumstances" of the parties to the UNFCCC, this flexibility in the interpretation of CBDR-RC is an important advantage of the equity calculator over other climate justice indices.

At the same time, thanks to its equity band, the calculator also sets a robust framework to make sure that the chosen parameters do not become unfair. This is clearly illustrated by the capacity indicator. Both the GDR framework (Baer et al. 2009, 1125) and the CERF (Holz, Kartha, and Athanasiou 2018, 124) tend to set the development threshold at $7,500 annually, or $20 per person per day, because this level "is close to the poverty line of industrialized countries, and to the lower bound of a 'consumer class' in developing countries" (Baer 2013, 64). In other words, it would be unfair to require people below this line to mitigate their emissions.[21]

But isn't it unrealistic, as Ewan Kingston (2016, 5–6) objects, to set such a development threshold "in a milieu that typically designates two dollars per day as the benchmark for global poverty?" While $20 a day can indeed be considered by some countries as too high to set the development threshold, $2 a day would definitely be too low. Approximately $16 per day "sets the point at which the classic plagues of poverty—malnutrition, high infant mortality, low educational attainment, high relative food expenditures—begin to disappear, or at least become exceptions to the rule" (Baer et al. 2009, 1125). Setting the development threshold below this line is therefore difficult to justify, especially if we add that this threshold is only based on per capita incomes, and not on other basic capabilities measured by the Human Development Index (HDI), such as education and life expectancy. Setting the development threshold at $2 per day would imply that each person with an income above this line should be expected to make sacrifices to contribute to the fight against climate change. The right to development, recognized by the IPCC as a fundamental normative dimension of any burden-sharing framework (Holz, Kartha, and Athanasiou 2018, 212), would be compromised. The minimum level of the equity band is therefore approximately $16 per day—in any case, not as low as $2 a day. Basic demands of justice should not be sacrificed in the name of realism.

To sum up, due to its flexibility in terms of the different parameters for measuring responsibility and capacity, the equity calculator has a relatively high degree of feasibility; at the same time, due to the limits the equity band puts on the choices of parameters, it avoids the risk of being insufficiently demanding and thereby guarantees a certain degree of fairness.

The Role of Civil Society

In addition to the vagueness of CBDR-RC, another concern one might raise is the role played by NGOs in the design of the equity calculator. For instance, Kingston (2016, 5) objects that the provenance of the Climate Equity Reference Calculator from "a group of NGOs that historically have often been highly critical of developed countries' positions in international affairs will reduce the review's global credibility." To reply to this objection, some considerations on the methodology of the CERF are in order.

Several NGOs, including Oxfam, WWF, and 350.org supported the original report proposing the equity calculator at COP21 (CERP 2015). However, the whole project is an initiative of EcoEquity and the Stockholm Environment Institute, two research organizations dealing with environmental issues. Holz, Kartha, and Athanasiou (2018) are affiliated with these two institutions. These three scholars included civil society organizations in the deliberative process that defined the effort-sharing parameters and equity band. This led to an "Equity Review coalition" that represented "a broad spectrum of perspectives and backgrounds, from large international environmental NGOs to Southern grassroots justice movements, from trade unions to development aid organizations to faith-based organizations" (Holz, Kartha, and Athanasiou 2018, 124). This "broad spectrum of perspectives" contributes both to the flexibility and the robustness of the operationalization of CBRD-RC. So, the response to Kingston's objection is that (1) the equity calculator was developed by scholars, not activists, even if Holz, Kartha, and Athanasiou have relied on the experience, expertise, and feedback of NGOs; and that (2) the calculator is supported by many different institutions, activists, and scholars, along with diplomats from BASIC countries (Bourban 2017, 17–18), and this actually increases the credibility of the equity calculator.

One could object that civil society does not have enough power to promote this kind of institutional reform in the fighting pit of international negotiation and so the reform is unrealistic or infeasible. The geopolitics of climate change is essentially a geopolitics of fossil energy dominated by China, Russia, and the United States, each determined to extract and use most of the remaining gas, oil, and coal (Aykut and Dahan 2015, 430–433). Together with fossil fuel energy lobbyists and the Gulf Cooperation Council (GCC), these three countries have indeed played a key role in slowing down or blocking progress in the fight against climate change. (For more on the role of fossil fuel companies and lobbies, please see Hossain et al. in this volume.)

To reply to this objection, it is crucial to stress that civil society has become a powerful actor in climate negotiations. It is to be expected that efforts aiming to close the equity gap would meet some form of resistance, but civil society has become a major counter-power to fossil fuel lobbyists and climate laggards, such as the United States and Saudi Arabia. Since the early 1990s, environmental NGOs have gained considerable financial and symbolic power. They have grown in size and weight, especially through their influence on decision-makers, diplomats, and economic actors. Over the years, together with influential activists such as Naomi Klein, Bill McKibben, and more recently Greta Thunberg, they have become members of a global movement for climate justice. For many years, a fruitful dialogue on climate justice has taken place between activists, academic research teams (such as the CERF), and diplomats and other policymakers. The climate justice movement is a catalyst for transformation. It is now constantly pushing states to make more ambitious commitments than those proposed in the first round of NDCs in 2015. Since democratic countries are sensitive to public opinion, civil society campaigns can put pressure on their governments (1) to better justify their mitigation targets, (2) to set more ambitious targets, and (3) to specify the measures they will use to achieve their objectives. Such pressures from relatively powerful civil society groups make the inclusion of an equity calculator in the climate regime and/or NDCs a more feasible institutional reform.

The feasible set constantly evolves because of shifting power relationships. There are certainly powerful economic and political actors opposed to the proposed institutional reform. But there are also strong counter-powers that would favor the inclusion of an equity calculator in the climate regime to compare NDCs' level of ambition and equity. These include civil society, developing (e.g., AOSIS) countries, emerging (e.g., BASIC) countries, and some developed country parties (e.g., EU).

The fact that Posner and Weisbach, the GCC, and the United States are surely against the inclusion of an equity calculator in NDCs and in the climate regime does not mean it is infeasible. First, there are major counter-powers to promote this kind of institutional reform to make climate policies fairer and more ambitious. Second, the United Nations Environment Programme (UNEP) does not necessarily need the agreement of all the parties to the UNFCCC to publish an *Equity Gap Report* or to add a new chapter on this topic in their *Emissions Gap Report*. Science and policy are interlinked in the climate regime, but scientific expertise should remain independent from political influence. Third, AOSIS, BASIC, and some EU countries can start including an equity calculator in their NDCs without having to wait for climate laggards to do the same. Indeed, as I stress in the next section, this could even push other countries to do the same, out of reputational concern.

To sum up, in what sense can civil society help to promote climate justice? It can support institutional reforms in collaboration with scholars, as the Equity Review Coalition illustrates very well. This kind of collaboration is necessary to push states to implement more ambitious climate policies. The reform proposal discussed in this chapter is insufficient on its own to guarantee that the equity gap will be substantially reduced. More modestly, my point is to contribute, with other projects common to climate justice scholars and members of civil society, to promoting climate justice in the climate regime. The idea is that a common and transparent normative framework will push states to propose fairer pledges when they have to revise their NDCs, as part of the five-year cycle set by the Paris Agreement. Other measures are necessary, for instance, to increase the likelihood that states will respect their mitigation pledges in their updated NDCs. Here, a key complementary reform would be to complement the pledge-and-review approach with a price-signal approach (Bourban Forthcoming-b).

Civil society is a major agent of change. So far, the key role of this political actor has been underexplored in the climate justice literature. Since civil society is strongly advocating in favor of climate justice, this is surprising, and the dominance of ideal theory in political philosophy may be a factor in this. The rise of nonideal approaches to climate justice means this is slowly starting to change. For instance, Caney (2016b, 23–24) states, "organizations can and should operate within this existing governance structure and exploit these review processes and mechanisms to hold governments to account." He takes the example of the Climate Action Tracker, another climate justice index: "It is valuable for NGOs to publish reports ranking parties in terms of their compliance with this, or to produce a league table of how each government is doing (both in terms of how ambitious its commitments are and to what extent it is matching them)." This approach represents "a non-ideal response to the existing international climate regime," because it "takes the existing architecture as a starting point, and seeks to find ways to improve it" (ibid.).[22]

Reputational Concerns

Now that I have stressed how the flexibility of the equity calculator and the counter-power of civil society contribute to the feasibility of the reform proposal, I would like to stress a third factor: the soft power of reputation. Countries that sign treaties have strong reasons to adhere to them for reasons of reputation. Reputation costs represent informal or symbolic sanctions "brought to bear on countries that fail to meet their own targets or to set ambitious goals" (Jacquet and Jamieson 2016, 645). When countries pledge their new mitigation targets every five years, other countries, and members of civil society will review their updated NDCs and hold them accountable. Informal sanctions or reputational costs include "naming and shaming," the threat or use of public opprobrium to affect reputation (Jacquet 2015). According to Kingston (2016, 2) "[o]ne of the missing elements in the Paris Agreement is a formal mechanism by which such reputational effects can be generated and amplified." The equity calculator could play precisely this role.

In a bottom-up regime of regularly reassessed national commitments, reputation is key. As we have already seen in the first round of NDCs, countries that committed to doing very little, such as the United States and Australia, endured criticism from other countries and civil society. But the most important element is not necessarily naming and shaming. More positively, the more that developed states make commitments that match their fair share to protect their reputation, the more likely other states are to agree to increase the ambition of their own commitments. If high-capacity and high-responsibility countries offer more equitable NDCs to improve their reputations, a virtuous circle may be set in motion, increasing the effectiveness of the climate regime. As the IPCC Working Group III (2014b, 5) stresses, "[t]he evidence suggests that outcomes seen as equitable can lead to more effective cooperation." Jacquet and Jamieson (2016, 645) add that while it is indeed "easier to avoid one's commitments if others are avoiding theirs," it also "becomes more difficult not to fulfil one's commitments if others are fulfilling theirs."

It is not necessary for all the parties to the UNFCCC to include an equity calculator in their updated NDCs at the same time for this kind of reputational concern to work. As long as some countries start to lead by example, others will be pushed to do the same. There are already several possible candidates. First, BASIC countries expressed their interests in the project at COP21 (Bourban 2017, 17–18). Second, taking the norm of CBDR-RC as a criterion for assessing national contributions, Axel and Katharina Michaelowa (2015) show that a growing number of emerging and developing countries with high population, and economic and GHG growth, are accepting new mitigation responsibilities. This is true of Indonesia and South Africa despite pressure from fossil fuel lobbies. Singapore and China are also becoming increasingly concerned with their national climate policies. Third, the EU is known for its historic leadership in climate negotiation, as well as for its insistence on equity as a central criterion for burden-sharing in terms of mitigation (Aykut and Dahan 2015, 216–250). Finally, AOSIS countries are famous for their "moral leadership" on setting the temperature rise limit at 1.5°C above pre-industrial levels, on promoting justice in climate adaptation finance, and on pushing for legally binding outcomes (de Águeda Corneloup and Mol 2014). They are probably among the most likely groups of countries that would agree to include some form of equity calculator in their NDCs.

In a bottom-up architecture where any attempt to impose a framework from above is doomed to failure, especially on powerful actors such as the United States, there might be some remaining concerns that the reform proposal will prove ultimately infeasible. Two replies could be made. First, as Crawford (2002, 103) highlights, ethical considerations "may change actors' conceptions of what is possible or what is desirable," and "can even result in actors reframing their 'interests,' or the order of their preferences." Ethical considerations can, for instance, increase the willingness of countries to lead by example. Within the UNFCCC, reputational concerns could be generated by a transparent normative framework that would increase the visibility of the ambition gap and allow easier comparison of each country's contribution in its updated NDC. By increasing visibility and comparability, the equity calculator could put the ambition gap on the political agenda of many democratic countries and play a role in elections.

Second, countries such as the United States could be motivated to take climate leadership seriously for a plurality of reasons: moral motivations can play a role, but amoral or strategic considerations such as domestic approval from citizens also certainly matter. According to Light (2017, 496), "[w]hether generated by geo-political self-interest or a sense of global responsibility, these kinds of larger reputational benefits for ambition will have positive benefits for the climate." The United States has a history of playing a role of leader on the international stage, for instance, in terms of technological innovation.[23] One way to lead by example here would be to direct efforts and public investments in the energy transition, which partially relies on technological innovation, especially in renewable energy (Bourban Forthcoming-a). In addition to reduction in GHG emissions, phasing out fossil fuels also leads to cleaner air and water, more permanent, full-time jobs, as well as energy security and independence (Jacobson et al. 2017). This could motivate many countries to seriously engage in the energy transition and agree to incorporate an equity calculator in their new, more ambitious pledges to measure their progress.

IV. CONCLUSION

A nonideal approach to climate justice seeks institutional reforms that are both fair and feasible, given relevant political constraints. It is highly fact-sensitive, drawing on empirical data on the functioning of the climate system and of institutions and international negotiations. Because of the scarce remaining global carbon budget, we have to philosophize in a situation of climate emergency. In such circumstances, closing the ambition gap between the demands of climate justice and existing mitigation policies means finding institutional reforms that start from where we are now and contribute to transitional improvements. The inclusion of a normative framework in the climate regime and NDCs based on a climate justice index, such as the one proposed by the CERF, is an important step in the transition between existing mitigation policies and the demands of principles of justice. Since the index is based on CBDR-RC, the cornerstone of the climate regime, and since CBDR-RC reflects basic demands of distributive climate justice, the proposal complies with the demands of fairness within the limits of the feasible set. Many other and complementary institutional reforms are also possible, and strongly needed.

Nonideal theory identifies obstacles to climate progress and considers how to respond to them in a way that takes into account relevant political feasibility constraints. The two main obstacles to reduce the ambition gap I investigated here are

(1) The vagueness of the norm of CBDR-RC: the parties to the UNFCCC nominally agreed to CBDR-RC but (a) it can be interpreted in multiple ways and (b) states can interpret it in ways that suit their self-interest.
(2) The power of political actors thwarting climate progress: major fossil fuel providers and consumers, such as China, Russia, and the United States,

together with fossil fuel energy lobbyists and the GCC, have systematically undermined progress in multilateral efforts to mitigate global emissions.

Regarding (1), I have stressed that the climate equity reference calculator is flexible enough to allow multiple countries to measure responsibility and capacity in the light of their different national circumstances, but, thanks to its equity band, it is also sufficiently robust to ensure that basic normative demands such as the PPP, the APP, and the right to development are all included in the operationalization of CBDR-RC. Since the Paris Agreement requires states to increase the ambition of their NDCs by taking CBDR-RC seriously, an equity calculator allowing a flexible operationalization of this norm is a precious tool in reducing the ambition gap between principles of climate justice and climate policies. The point is not to aim for perfect justice but to allow for nonideal justice to guide the most urgent climate policies in our context of climate emergency.

Regarding (2), I have stressed the crucial role played by civil society in promoting this tool. Civil society is a key counterpower to energy lobbies and climate laggards. I have also highlighted the power of norms and values in international negotiations by drawing on constructivist international relations theory, which claims that ethical arguments can change actors' conception of what is possible and desirable. Finally, I have explained that ethical considerations have played a crucial role in the creation and development of the climate regime. Normative concerns are widespread in its most recent outcomes, such as the Paris Agreement, the IPCC Fifth Assessment Report, and the IPCC Special Report on Global Warming of 1.5°C. It is true that so far, these concerns have been downplayed because of the short-term economic thinking of fossil fuel providers and consumers. It is time that the numerous ethical considerations that have shaped the climate regime became more influential in policymaking. For this to happen, normative political theorists need to develop more relevant action-guiding considerations, but decision-makers also need to recognize that climate change cannot be separated from climate justice. A nonideal approach to climate justice is a strong way to meet this interdisciplinary challenge.[24]

NOTES

1. See several contributions in the two following anthologies: Gardiner et al. (2010); Harris (2016). The distinction between burden-sharing justice and harm avoidance justice has been developed by Caney (2014).

2. For a recent volume on the topic of historical responsibility, see Meyer and Sanklecha (2017).

3. According to a recent study by Comyn-Platt et al. (2018), if we include in the calculation methane emissions from natural wetlands and carbon release from thawing permafrost, the carbon budget for stabilization at 1.5°C is reduced by 9–15 percent (25–38 GtC), and by 6–10 percent (33–52 GtC) for 2°C stabilization.

4. Others have also started to pave the way for such an approach: Bell (2013); Caney (2016a, 2016b); Heyward and Roser (2016); Gajevic Sayegh (2018); Brandstedt (2019).

5. For an overview of the three dominant interpretations in the literature, see Valentini (2012).

6. I only deal here with responsibilities for mitigation. Developed countries however also have responsibilities in terms of help with adaptation, and in terms of compensation and reparation for loss and damage. On the topic of adaptation finance justice, see Baatz and Bourban (2019); on the topic of justice for loss and damage, see Wallimann-Helmer (2019). Following Caney (2005), I consider that the principles that lead to a fair allocation of mitigation burdens are the same as those that lead to a fair allocation of adaptation burdens and compensation burdens: see Bourban (2018, 111–118). Mitigation, adaptation, and compensation policies are complementary, but as Shue (2016) stresses, mitigation remains the "first imperative."

7. Note that the two tasks of a nonideal approach are complementary because (a) implementing fair and effective climate policies requires political and economic actors to be motivated to adopt them, and (b) developing institutional reforms can lead to discussions about the motivations driving actors to comply with certain principles.

8. While (2) and (3) are stressed by Valentini (2012), (1) is explained by Gajevic Sayegh (2018, 407), according to whom "in order to guide action, our ideal principles of climate justice need to be reformulated in the light of real-world considerations, which we only obtain by integrating the relevant empirical work on the matter."

9. For reflections on different types and degrees of feasibility, I draw on Gilabert (2012, 117–122).

10. Some argue that betting on negative emissions could allow us to expand our scarce global carbon budget and render current emissions overshooting by 2050 compatible with stringent warming targets by 2100. Scenarios with a heavy reliance on carbon dioxide removal (CDR) based on a large-scale deployment of negative emissions technologies (NETs) however raise too many ethical, ecological, and political problems, as shown by the emerging field of the ethics of NETs (Shue 2017, Dooley and Kartha 2018, Lenzi 2018, Bourban Forthcoming-a).

11. As I am making the final revisions to this chapter, a new event, particularly relevant in the context of climate negotiations, could contribute to change the feasible set: the election of Joe Biden as President-elect of the United States. Biden has promised to rejoin the Paris Agreement and to substantially invest in low-carbon technology (especially with a $1.7tn investment plan in a green recovery from the COVID crisis) (Harvey 2020). Several challenges are however still in the way, especially the following two: (1) the election of Biden as President of the United States still needs to be officially confirmed; (2) even if he does become President, if the United States rejoins the Paris Agreement, and if an updated U.S. NDC is submitted

for COP26, Biden will face strong domestic opposition from the Senate, the Supreme Court, and business interests. That being said, the fact that Biden seems committed to take strong climate action, together with the fact that multiple states, cities, and local governments already pledged to take action compatible with the goal set by the Paris Agreement (the "We Are Still In" coalition), shows that the degree of feasibility of more ambitious mitigation measures at the national and international levels can increase in the coming years (CAT 2020).

12. Posner and Weisbach (2010, 86) argue indeed that "an optimal climate treaty would probably not require side payments to poor countries. It could well require side payments to rich countries like the United States and rising countries like China, and indeed possibly from very poor countries which are extremely vulnerable to climate change—such as Bangladesh."

13. Andrew Light (2017, 491) explains that this consensus rule represents an important moral dimension of the climate regime: it is a "principle of procedural justice" that recognizes "the common risk of all parties to the threat of climate change regardless of their size or political power." Another fundamental moral dimension of climate diplomacy addressed in detail below is the principle of distributive justice embodied in the norm of CBDR-RC.

14. The Fifth IPCC Assessment Report was the first to include two philosophers as lead authors (John Broome and Lukas Meyer). For more about the role of philosophers in the IPCC, see Broome (2020).

15. While the Paris Agreement does not require parties to implement their NDCs, it requires, as a legal obligation, each party to communicate a successive NDC every five years (Bodansky 2016).

16. For a synthesis of burden-sharing frameworks in the literature, see Working Group III (2014a, 315–321). For an interpretation of the early results and normative implications of the first report published by the Climate equity reference project (CERP 2015), see Bourban (2017).

17. Alternative levels of ambition could be proposed on the basis of realism (less-demanding goals could be easier to achieve), but they would be unreasonable in the light of recent scientific findings: any global carbon budget aiming for less than 66 percent chance of staying within 2°C would increase the risks of triggering a cascade of tipping points in the climate system, possibly pushing the whole Earth system irreversibly onto a "hothouse Earth pathway" (Steffen et al. 2018).

18. I explain in the next section how the authors of the CERF set this equity band by replying to some objections.

19. For instance, Éloi Laurent (2020, 106–110) measures responsibility with consumption emissions per capita (averaged over 1990–2012), capacity with the level of development according to the HDI (also averaged over 1990–2012), and also adds the projected population increase until 2050. He finds that India should be allocated 78 billion tons of CO_2 to be emitted by 2050, Brazil 62 billion tons, South Africa 51 billion tons, and China 48 billion tons. In contrast, the United States owes 17 billion tons of CO_2 to the rest of the world, Canada 9 billion tons, Germany 2 billion tons, and Japan 1 billion tons. In other words, developed countries have a "negative carbon budget" and have to pay "by investing in carbon sinks or by transferring technology

and/or financing to accelerate emission reductions in carbon positive carbon budget countries." Although these indicators of responsibility and capacity are quite different from the CERF, the main results of the operationalization of the norm of CBDR-RC are therefore similar: while developing countries have a low degree of responsibility to mitigate global emissions, developed countries have a high degree of responsibility to do so (which includes a responsibility to contribute to mitigation efforts elsewhere through climate finance), and emerging countries such as China, South Africa, and Brazil have an intermediate level of responsibility.

20. I thank Simon Caney for our discussions on this point.

21. A realist-minded objector could argue here that such a development threshold could actually seem unfair to some developed countries because it would let too many people in developing countries off the hook and would put too much responsibility on richer people. Even if there is some truth in this objection, it remains substantially more unfair to put more responsibility on poor people (who have to live with less than $20 per day) than on rich people, for the simple reason that, as Shue (1992, 397) put it vividly, "whatever justice may positively require, it does not permit that poor nations be told to sell their blankets in order that rich nations may keep their jewellery."

22. It is encouraging to see that some climate justice scholars are increasingly engaged in public activities beyond academia. Simon Caney provides a good example of such public engagement, having worked closely with members of public bodies, NGOs, human rights organizations, research and policy institutes, and international institutions on ethical issues raised by climate change. His contributions can be found on his website: https://simoncaney.weebly.com/public-engagement.html, accessed 14.04.2020.

23. For instance, it is not a coincidence that many geoengineering projects started and are still under developed in the United States.

24. For their very helpful comments on how to improve this chapter, I am grateful to Lisa Broussois and Simon Caney, and to the editors of this volume, Corey Katz and Sarah Kenehan. I would also like to thank ProClim—Forum for Climate and Global Change, which enabled me to take part in COP21 in Paris as an observer, where I started to work on considerations developed in this chapter. I gratefully acknowledge financial support from the Swiss National Science Foundation (grant P400PG_190981).

REFERENCES

Allen, M.R., O.P. Dube, W. Solecki, F. Aragón-Durand, S. Humphreys W. Cramer, M. Kainuma, J. Kala, N. Mahowald, Y. Mulugetta, R. Perez, M. Wairiu, and K. Zickfeld. 2018. "2018: Framing and Context." In *Global Warming of 1.5°C. An IPCC Special Report on the impacts of global warming of 1.5°C above pre-industrial levels and related global greenhouse gas emission pathways, in the context of strengthening the global response to the threat of climate change, sustainable development, and efforts to eradicate poverty*, edited by V. Masson-Delmotte, P. Zhai, H.-O. Pörtner, D. Roberts, J. Skea, P.R. Shukla, A. Pirani, W.

Moufouma-Okia, C. Péan, R. Pidcock, S. Connors, J.B.R. Matthews, Y. Chen, X. Zhou, M.I. Gomis, E. Lonnoy, T. Maycock, M. Tignor, and T. Waterfield, 49–81. Geneva: World Meteorological Organization.

André, Pierre, and Michel Bourban. 2016. "Éthique et justice climatique : entre motivations morales et amorales." *Les ateliers de l'éthique/The Ethics Forum* 11(2–3): 4–27.

Aubertin, Catherine, Michel Damian, Michel Magny, Claude Millier, Jacques Theys, and Sébastien Treyer. 2015. "Introduction. Les enjeux de la conférence de Paris. Penser autrement la question climatique." *Natures Sciences Sociétés* Supplément(Supp. 3): 3–5.

Aykut, Stefan C., and Amy Dahan. 2015. *Gouverner le climat ? Vingt ans de négociations internationales*. Paris: Presses de Sciences Po.

Baatz, Christian. 2018. "Climate Adaptation Finance and Justice. A Criteria-Based Assessment of Policy Instruments." *Analyse & Kritik* 40(1): 73–105.

Baatz, Christian, and Michel Bourban. 2019. "Distributing Scarce Climate Adaptation Finance Across SIDS: Effectiveness, Not Efficiency." In *Dealing with Climate Change on Small Islands: Towards Effective and Sustainable Adaptation?*, edited by Carola Klöck and Michael Fink, 77–99. Göttingen: Göttingen University Press.

Baer, Paul. 2013. "The Greenhouse Development Rights Framework for Global Burden Sharing: Reflection on Principles and Prospects." *Wiley Interdisciplinary Reviews: Climate Change* 4(1): 61–71.

Baer, Paul, Sivan Kartha, Tom Athanasiou, and Eric Kemp-Benedict. 2009. "The Greenhouse Development Rights Framework: Drawing Attention to Inequality within Nations in the Global Climate Policy Debate." *Development and Change* 40(6): 1121–1138.

Bell, Derek. 2013. "How Should We Think about Climate Justice?" *Environmental Ethics* 35(2): 189–208.

Bodansky, Daniel. 2016. "The Legal Character of the Paris Agreement." *Review of European, Comparative & International Environmental Law* 25(2): 142–150.

Boffey, Daniel, and Jennifer Rankin. 2020. "EU Leaders Seal Deal on Spending and €750bn Covid-19 Recovery Plans." The Guardian, Last Modified 21.07.2020. https://www.theguardian.com/world/2020/jul/20/macron-seeks-end-acrimony-eu-summit-enters-fourth-day.

Bourban, Michel. 2014. "Climate Change, Human Rights and the Problem of Motivation." *De Ethica* 1(1): 37–52.

Bourban, Michel. 2017. "Justice climatique et négociations internationales." *Négociations* 27(1): 7–22.

Bourban, Michel. 2018. *Penser la justice climatique. Devoirs et politiques*. Paris: PUF.

Bourban, Michel. Forthcoming-a. "Ethics, Energy Transition, and Ecological Citizenship." In *Encyclopedia Of Comprehensive Renewable Energy*, edited by Ali Sayigh. Amsterdam/San Diego: Elsevier.

Bourban, Michel. Forthcoming-b. "Promoting Justice in Climate Policies." In *Routledge Handbook on the Political Economy of the Environment*, edited by Éloi Laurent, Rebecca Ray and Klara Zwickl. New York: Routledge.

Brandstedt, Eric. 2019. "Non-ideal Climate Justice." *Critical Review of International Social and Political Philosophy* 22(2): 221–234.
Broome, John. 2020. "Philosophy in The IPCC." In *A Guide to Field Philosophy: Case Studies and Practical Strategies*, edited by Evelyn Brister and Robert Frodeman. New York: Routledge.
Caney, Simon. 2005. "Cosmopolitan Justice, Responsibility, and Global Climate Change." *Leiden Journal of International Law* 18(4): 747–775.
Caney, Simon. 2010. "Climate Change and the Duties of the Advantaged." *Critical Review of International Social and Political Philosophy* 13(1): 203–228.
Caney, Simon. 2014. "Two Kinds of Climate Justice: Avoiding Harm and Sharing Burdens." *Journal of Political Philosophy* 22(2): 125–149.
Caney, Simon. 2016a. "Climate Change and Non-Ideal Theory: Six Ways of Responding to Non-Compliance." In *Climate Justice in a Non-Ideal World*, edited by Clare Heyward and Dominic Roser, 21–42. Oxford: Oxford University Press.
Caney, Simon. 2016b. "The Struggle for Climate Justice in a Non-Ideal World." *Midwest Studies In Philosophy* 40(1): 9–26.
CAT. 2020. "Biden's Election Could Bring a Tipping Point Putting Paris Agreement 1.5 Degree Limit "Within Striking Distance"." Last Modified 07.11.2020. https://climateactiontracker.org/press/bidens-election-could-bring-a-tipping-point-putting-paris-agreement-15-degree-limit-within-striking-distance/.
CERP. 2015. *Fair Shares: A Civil Society Equity Review of INDCs: Summary*. Manila, London, Cape Town, Washington: Climate Equity Reference Project.
Chung, Ryoa. 2012. "Théories idéale et non idéale." In *Éthique des relations internationales*, edited by Ryoa Chung and Jean-Baptiste Jeangène Vilmer, 63–91. Paris: PUF.
Coady, David, Ian Parry, Louis Sears, and Baoping Shang. 2017. "How Large Are Global Fossil Fuel Subsidies?" *World Development* 91: 11–27.
Comyn-Platt, Edward, Garry Hayman, Chris Huntingford, Sarah E. Chadburn, Eleanor J. Burke, Anna B. Harper, William J. Collins, Christopher P. Webber, Tom Powell, Peter M. Cox, Nicola Gedney, and Stephen Sitch. 2018. "Carbon budgets for 1.5 and 2 °C targets lowered by natural wetland and permafrost feedbacks." *Nature Geoscience* 11(8): 568–573.
Crawford, Neta C. 2002. *Argument and Change in World Politics: Ethics, Decolonization, and Humanitarian Intervention*. Cambridge: Cambridge University Press.
de Águeda Corneloup, Inés, and Arthur P. J. Mol. 2014. "Small Island Developing States and International Climate Change Negotiations: The Power of Moral "Leadership"." *International Environmental Agreements: Politics, Law and Economics* 14(3): 281-297.
Dooley, Kate, and Sivan Kartha. 2018. "Land-Based Negative Emissions: Risks for Climate Mitigation and Impacts on Sustainable Development." *International Environmental Agreements: Politics, Law and Economics* 18(1): 79–98.
Gajevic Sayegh, Alexandre. 2018. "Justice in a Non-Ideal World: the Case of Climate Change." *Critical Review of International Social and Political Philosophy* 21(4): 407–432.

Gardiner, Stephen M., Simon Caney, Dale Jamieson, and Henry Shue. 2010. *Climate Ethics: Essential Readings*. Oxford: Oxford University Press.

Gardiner, Stephen M. 2011. *A Perfect Moral Storm : The Ethical Tragedy of Climate Change*. Oxford: Oxford University Press.

Gardiner, Stephen M., and David A. Weisbach. 2016. *Debating Climate Ethics*. New York: Oxford University Press.

Gilabert, Pablo. 2012. *From Global Poverty to Global Equality: A Philosophical Exploration*. Oxford: Oxford University Press.

Hansen, James E. 2009. *Storms of my Grandchildren: The Truth about the Coming Climate Catastrophe and our Last Chance to Save Humanity*. London: Bloomsbury.

Harris, Paul G. 2016. *Ethics, Environmental Justice and Climate Change*. Cheltenham/Northampton: Edward Elgar.

Harvey, Fiona. 2020. "Joe Biden Could Bring Paris Climate Goals 'Within Striking Distance'." The Guardian, Last Modified 08.11.2020. https://www.theguardian.com/us-news/2020/nov/08/joe-biden-paris-climate-goals-0-1c.

Heyward, Clare, and Jörgen Ödalen. 2016. "A Free Movement Passport for the Territorially Dispossessed." In *Climate Justice in a Non-Ideal World*, edited by Clare Heyward and Dominic Roser, 208–226. Oxford: Oxford University Press.

Heyward, Clare, and Dominic Roser. 2016. *Climate Justice in a Non-Ideal World*. Oxford: Oxford University Press.

Holz, Christian, Sivan Kartha, and Tom Athanasiou. 2018. "Fairly Sharing 1.5: National Fair Shares of a 1.5 °C-Compliant Global Mitigation Effort." *International Environmental Agreements: Politics, Law and Economics* 18(1): 117–134.

IPCC (Intergovernmental Panel on Climate Change). 2018. "Summary for Policymakers." In *Global Warming of 1.5°C. An IPCC Special Report on the impacts of global warming of 1.5°C above pre-industrial levels and related global greenhouse gas emission pathways, in the context of strengthening the global response to the threat of climate change, sustainable development, and efforts to eradicate poverty*, edited by V. Masson-Delmotte, P. Zhai, H.-O. Pörtner, D. Roberts, J. Skea, P.R. Shukla, A. Pirani, W.

Jacobson, Mark Z., Mark A. Delucchi, Zack A. F. Bauer, Savannah C. Goodman, William E. Chapman, Mary A. Cameron, Cedric Bozonnat, Liat Chobadi, Hailey A. Clonts, Peter Enevoldsen, Jenny R. Erwin, Simone N. Fobi, Owen K. Goldstrom, Eleanor M. Hennessy, Jingyi Liu, Jonathan Lo, Clayton B. Meyer, Sean B. Morris, Kevin R. Moy, Patrick L. O'Neill, Ivalin Petkov, Stephanie Redfern, Robin Schucker, Michael A. Sontag, Jingfan Wang, Eric Weiner, and Alexander S. Yachanin. 2017. "100% Clean and Renewable Wind, Water, and Sunlight All-Sector Energy Roadmaps for 139 Countries of the World." *Joule* 1(1): 108–121. doi:

Jacquet, Jennifer. 2015. *Is Shame Necessary? New Uses for an Old Tool*. New York: Pantheon.

Jacquet, Jennifer, and Dale Jamieson. 2016. "Soft but Significant Power in the Paris Agreement." *Nature Climate Change* 6(7): 643–646.

Kingston, Ewan. 2016. "Clustering Countries, Changing Climates: An NGO Review to Close the Ambition Gap." *Ethics & International Affairs* (Online): 1–10.

Laurent, Éloi. 2020. *The New Environmental Economics*. Cambridge/Medford: Polity.
Lenzi, Dominic. 2018. "The Ethics of Negative Emissions." *Global Sustainability* 1: 1–8.
Light, Andrew. 2017. "Climate Diplomacy." In *The Oxford Handbook of Environmental Ethics*, edited by Stephen Gardiner and Allen Thompson, 487–500. Oxford: Oxford University Press.
Light, Andrew, and Gwynne Taraska. 2016. "A Responsible Path: Enhancing Action on Short-Lived Climate Pollutants." In *Climate Justice in a Non-Ideal World*, edited by Clare Heyward and Dominic Roser, 169–188. Oxford: Oxford University Press.
Maltais, Aaron. 2016. "A Climate of Disorder: What to Do About the Obstacles to Effective Climate Politics." In *Climate Justice in a Non-Ideal World*, edited by Clare Heyward and Dominic Roser, 43–63. Oxford: Oxford University Press.
McKelvey, Tara. 2020. "US election 2020: Congressional Virus Relief Talks Grind On." BBC News, Last Modified 21.10.2020. https://www.bbc.co.uk/news/election-us-2020-54619918.
Meyer, Lukas H., and Pranay Sanklecha. 2017. *Climate Justice and Historical Emissions*. Cambridge: Cambridge University Press.
Michaelowa, Axel, and Katharina Michaelowa. 2015. "Do Rapidly Developing Countries Take Up New Responsibilities for Climate Change Mitigation?" *Climatic Change* 133(3): 499–510.
Ohlin, Jens D. 2015. *The Assault on International Law*. Oxford/New York: Oxford University Press.
Page, Edward A. 2011. "Climatic Justice and the Fair Distribution of Atmospheric Burdens: A Conjunctive Account." *The Monist* 94(3): 412–432.
Page, Edward A., and Clare Heyward. 2016. "Compensating for Climate Change Loss and Damage." *Political Studies* 65(2): 356–372.
Posner, Eric A., and David Weisbach. 2010. *Climate Change Justice*. Princeton: Princeton University Press.
Rawls, John. 1999. *A Theory of Justice: Revised Edition*. Cambridge: The Belknap Press of Harvard University Press.
Roser, Dominic. 2015. "Climate Justice in the Straitjacket of Feasibility." In *The Politics of Sustainability: Philosophical Perspectives*, edited by Dieter Birnbacher and May Thorseth, 71–91. New York/London: Routledge.
Shue, Henry. 1992. "The unavoidability of justice." In *The International Politics of the Environment: Actors, Interests, and Institutions*, edited by Andrew Hurrell and Benedict Kingsbury, 373–397. Oxford: Clarendon Press.
Shue, Henry. 1996. "Environmental Change and the Varieties of Justice." In *Earthly Goods*, edited by Fen Osler Hampson and Judith Reppy, 9–29. Cornell University Press.
Shue, Henry. 2015. "Historical Responsibility, Harm Prohibition, and Preservation Requirement: Core Practical Convergence on Climate Change." *Moral Philosophy and Politics* 2(1): 7–31.
Shue, Henry. 2016. "Mitigation: First Imperative of Environmental Ethics." In *The Oxford Handbook of Environmental Ethics*, edited by Stephen Gardiner and Allen Thompson, 465–473. Oxford: Oxford University Press.

Shue, Henry. 2017. "Climate Dreaming: Negative Emissions, Risk Transfer, and Irreversibility." *Journal of Human Rights and the Environment* 8(2): 203–216.

Steffen, Will, Johan Rockström, Katherine Richardson, Timothy M. Lenton, Carl Folke, Diana Liverman, Colin P. Summerhayes, Anthony D. Barnosky, Sarah E. Cornell, Michel Crucifix, Jonathan F. Donges, Ingo Fetzer, Steven J. Lade, Marten Scheffer, Ricarda Winkelmann, and Hans Joachim Schellnhuber. 2018. "Trajectories of the Earth System in the Anthropocene." *Proceedings of the National Academy of Sciences* 115(33): 8252–8259.

UNEP (United Nations Environment Programme). 2018. The Adaptation Gap Report 2018. Nairobi: United Nations Environment Programme

UNEP (United Nations Environment Programme). 2019. Emissions Gap Report 2019. Nairobi: United Nations Environment Programme.

UNFCCC (United Nations Framework Convention on Climate Change). 1992. United Nations Framework Convention on Climate Change. Document FCCC/INFORMAL/84. New York.

UNFCCC (United Nations Framework Convention on Climate Change). 2015. Adoption of the Paris Agreement. Decision 1/CP.21. Document FCCC/CP/2015/10/Add.1. Paris.

Valentini, Laura. 2012. "Ideal vs. Non-ideal Theory: A Conceptual Map." *Philosophy Compass* 7(9): 654–664.

Vanderheiden, Steve. 2016. "Climate Justice Beyond International Burden Sharing." *Midwest Studies In Philosophy* 40(1): 27–42.

Wallimann-Helmer, Ivo. 2019. "The Ethical Challenges in the Context of Climate Loss and Damage." In *Loss and Damage from Climate Change. Climate Risk Management, Policy and Governance*, edited by R. Mechler, L. Bouwer, T. Schinko, S. Surminski and Linnerooth-Bayer J., 39–62. Cham: Springer.

Working Group III to the Fifth Assessment Report of the Intergovernmental Panel on Climate Change. 2014a. *Climate Change 2014: Mitigation of Climate Change*. Cambridge: Cambridge University Press.

Working Group III to the Fifth Assessment Report of the Intergovernmental Panel on Climate Change. 2014b. "Summary for Policymakers." In *Climate Change 2014: Mitigation of Climate Change*. Cambridge: Cambridge University Press.

Chapter 4

International Law as a Basis for a Feasible Ability-to-Pay Principle

Ewan Kingston

Many philosophers working on climate change welcome the idea that principles of climate justice they develop will require the bulk of the burden of climate action to be borne by developed countries. However, principles of climate justice that are politically feasible should not just merely be acceptable to other fellow travelers but also be potentially persuasive to those who have reason to oppose the enactment of those principles. Consider two groups of such political opponents. One includes "reactionary citizens" who might take issue with climate justice in ways it is often conceived: as a species of global or intergenerational justice. For instance, they might suspect that most moral commitments are limited by state boundaries, or be skeptical that we have substantive duties to future people, or believe that those countries with current wealth primarily earned it and thus deserve it. The other group includes "constrained policymakers." These are policymakers who are sympathetic to progressive action on climate, but face constraints of public opinion from reactionary citizens (if an elected official or political appointee) or imposed by a primary duty to directly serve the medium-term interests of the state that employs them (if a career bureaucrat).

Faced with these political opponents, proponents of climate justice should consider how politically feasible different principles of climate justice are. I focus in this chapter on the political feasibility of an "ability-to-pay principle (APP)" as a proposal for dividing the burdens of past emissions and emissions from the global poor. I argue that a formulation of an APP with a voluntarist scope, restricted only to agreed-upon collective goals, is significantly more politically feasible than one with a preventative scope, which focuses on preventing a very bad outcome. A voluntarist APP can much more clearly be justified to both constrained policymakers and reactionary citizens as a solution to a collective action problem among states that have already bound themselves to collectively solving the problem. A voluntarist principle also has a stronger precedent as a principle of international law. Thus, those philosophers concerned with the political feasibility of an APP should carefully consider using a voluntarist rather than a preventative formulation.

The chapter proceeds as follows. First, I outline what I mean by "political feasibility" and distinguish it from other conceptions of feasibility. Second, I explore the different rationales for a preventative APP. Third, I introduce my voluntarist formulation of the APP. I then reply to objections to that formulation, arguing that the drawbacks of a voluntarist formulation are not as dire as they might appear.

I. ON POLITICAL FEASIBILITY

Feasibility, as analyzed by Juha Räikkä (1998), has at least two senses. "Theoretical feasibility" is a (potential) constraint on any political *theory* based on how societies could be organized. These constraints in turn might range from logical impossibility (a constraint on a theory that everyone should earn above the average wage)[1] or physical impossibility (a constraint on a theory that demands that each person receive an unlimited amount of energy or goods). More controversially, such constraints might include the standard motivational equipment humans have (Wiens 2016), the stock of resources needed to bring about a number of specific states of affairs (Wiens 2015), or the actual or perceived moral costs of moving from one overall structure to another (Buchanan 2002; Räikkä 1998). This notion of theoretical feasibility and how it relates to political theory and political philosophy (understood typically in terms of broad theories of governance or distribution) has received a lot of attention (Southwood 2018; Lawford-Smith 2013; Gilabert and Lawford-Smith 2012). What most parties seem to agree on is that the debate about theoretical feasibility typically concerns which real-world constraints should figure in our normative *thinking*. It does not concern something more practical, like the claims one might utter as advice or exhortation, for instance (Southwood 2018; Räikkä 1998).

Theoretical feasibility is not the only kind of feasibility that those concerned with climate justice might be interested in. Some concerned with climate justice might have reason to aim for feasibility in an even more conservative sense. Alongside theoretical feasibility, there is also what I will call (following Räikkä) "political feasibility." As Räikkä points out, it is perfectly natural for those interested in the actual "everyday politics" of concrete policies and institutions to take on board different (or stronger) feasibility constraints than those offered by even the most theoretical feasibility-conscious political theorists. For instance, it is common for political actors, such as a constrained policymaker, to dismiss a proposal as relatively infeasible if it (a) is not able to be brought about relatively quickly or (b) if it runs counter to the current beliefs of numerous and powerful constituencies. These dismissals assume nothing in particular about humans' motivational equipment or the costs of transitions.[2] In this chapter, I am interested in *political* feasibility, primarily in sense (b), the degree to which it can cohere with numerous and powerful constituencies. Indeed, I will refer to this scalar quality of being able to be consistent with the beliefs of numerous and powerful constituencies as "political feasibility." Political feasibility seems to be an appropriate quality to consider when we are trying to not just think but also prescribe what actors should do now, all things considered.[3] And given the urgency of the climate problem and the repeated calls for rapid action, focusing on what specific actors should do now seems highly relevant for the pursuit of climate justice.

Why should a *philosopher* be concerned with political feasibility? Is it not the role of a philosopher to hold actors to a *currently* politically infeasible standard? To continue to speak truth to power, even when it runs counter to widespread or powerful prevailing political attitudes? This is the model of the philosopher as a norm entrepreneur, or less optimistically, a Cassandra.[4] There is certainly a role for philosophers to put forward proposals that run counter to the political orthodoxy of their time. Indeed, demanding that constrained bureaucrats deliver courageous and just leadership, or that reactionary citizens accept the demands of a progressive, cosmopolitan view of justice, might be a way to spur at least an improvement from current unjust arrangements.

Taking the role of the uncompromising radical, however, can come with risks. A focus on highly politically infeasible ideals might sow division among otherwise powerful pragmatic coalitions. Further, disregarding political feasibility can leave one open to charges of irrelevance or ignorance and reduce any possible positive impact one might have on constrained policymakers or reactionary citizens.[5]

The model of the philosopher as pragmatist in public life might be less familiar to us than that of the philosopher as visionary. In these times of specialization, it's not likely that many philosophers can be deeply involved in policymaking (and when they try, they often admit to failing spectacularly).[6] But they can attempt to specifically develop arguments that might be ready out-of-the-box to assist policymakers, arguments that take widespread or powerful politically conservative opinions about global justice or international fairness as a constraint, not a target or an irrelevancy. Those arguments might help policymakers who are sympathetic to the cause of more radical climate justice to defend themselves against backlash from such constituencies. They might also be persuasive to those constituencies themselves and nudge their opinion in the direction of more radical visions of climate justice.[7]

In what follows I introduce the preventative formulation of the APP and argue it scores low on political feasibility. I then introduce my voluntarist version, which avoids the problems of the preventative formulation and can be justified to the political opponents mentioned above. I then defend this politically feasible voluntarist version, from two objections and a worry, arguing that in its application, it is not as different from the preventative formulation as one might think.

II. PREVENTATIVE ABILITY-TO-PAY PRINCIPLE

The APP, roughly put, dictates that richer states have a duty to do significantly more to help humanity reduce global emissions, adapt to climate changes, and perhaps compensate[8] unavoidable damages from climate change than poorer states. Notice, in this formulation, and in my voluntarist formulation that follows, *states* are the primary duty-bearers. There is a practical reason for this. The most plausible pathways to limiting warming to "well below 2°C" involve (i) electrification of manufacturing and transport systems, (ii) decarbonization of electricity generation systems, and possibly (iii) the deployment of large-scale negative emissions technology.[9] Each of (i)–(iii) are transformations that states are best placed to primarily bring about. Plausibly, in most states, (ii)—the decarbonization of the electricity generation system—is something that *requires* strong state action to bring about, given the huge infrastructure and national nature of most electricity systems. Therefore, a principle of burden-sharing that applies directly to states is one that directly applies to the relevant kind of actor to respond to climate change.[10]

The main aspect of the APP principle I want to explore is the "scope" of the principle. By scope I mean that distributive task which triggers the application of the principle. So if the APP is being proposed as a principle of distribution for some task(s) relating to climate change then the scope of the APP determines which particular task(s) it applies to. To ground this discussion, I lay out an APP principle with what I think is the default version of its scope—the version much writing in climate ethics implicitly gestures toward:

Preventative APP: The collective burden of *preventing a very bad outcome* should (where no better principle applies) be borne in relation to a state's ability to pay.

Notice three things about the principle. First, it refers to "preventing a very bad outcome." This is left unspecific. The idea is that there is a serious problem that *someone* has to deal with otherwise the outcome would be very bad. Whether that badness should be spelled out in terms of disutility, rights violations, or something else is secondary to the acknowledgment that there is a very bad outcome looming under business-as-usual that should be prevented if at all possible.[11] The second point is that the principle has explicit reference to its own low lexical priority. The preventative APP acknowledges that other principles which look to who is contributing to or benefiting from the very bad outcome might be more important. Indeed, several theorists see the APP as a way to determine shares of that part of the burden of climate change mitigation, adaptation (and compensation) that ought not be allocated, or cannot be allocated, via a principle with lexical priority such as a polluter pays principle (PPP). This "climate change remainder" (Caney 2010) is understood to result from greenhouse gases that have been produced (i) under conditions of ignorance about their effects (Kingston 2014), (ii) by people who are now dead (Caney 2010), or (iii) as a result of pursuing basic subsistence (Shue 1993; Caney 2010). The bad outcome of letting the burdens of this remainder fall just where they may calls for a principle of allocation beyond the PPP (Mittiga 2019; Caney 2010).[12] Third, as with the exact scope, the preventative APP is also unspecific about the *metric* of a state's ability to pay. There are numerous suggestions for measuring ability to pay in the climate case.[13] Some can be progressive. For instance, they can exclude incomes of individuals below a "subsistence" threshold and heavily weight incomes above a "luxury" level.[14] Alternatively, a measurement of state ability can be regressive, by correlating it directly to Gross National Income (Höhne, Den Elzen, and Escalante 2014). Moreover, a quick glance at the websites of the different public "fairness comparisons" of states' NDCs show that the choice of measurement metric makes a big difference to how burdens are ultimately divided.[15] (For more on fairness in relation to NDCs, please see Bourban in this volume.) This is true even under division schemes that incorporate both an APP and a PPP (Kingston 2020). I do not go into this problem here because it arises at a second stage in the discussion once an interlocutor has accepted that ability to pay is a relevant factor in the division of climate burdens. (For a treatment of the complex political feasibility issues at play in choosing an ability to pay metric, see Kingston 2020.)

The preventative APP is an assertion rather than an argument. If someone was unconvinced that ability to pay should play a role in dividing the burdens of climate change, specifically the remainder, the preventative APP gives them no further reason to accept it. Climate ethicists have, however, provided arguments for APP that seem tailor-made to justify the preventative APP. One way to assess these arguments is to try to determine whether they are successful or sound in relation to the intuitions of moral theorists. But another way to assess them, one that is highly relevant for political feasibility, is to see if they could be justified to those citizens that might be skeptical of those intuitions (Moellendorf 2016; Rawls 1999), or justified to policymakers constrained by either the beliefs of such citizens or institutional frameworks that enshrine a more conservative outlook on world affairs, such as a duty to serve the material self-interest of one's state. Thus, we can ask to what extent could the extant arguments for the preventative APP influence political opponents such as reactionary citizens or constrained policymakers?

Domestic Comparison

One kind of argument for the preventative APP proceeds by making an analogy to principles that might regulate intrastate relations between compatriots. For instance, Simon Caney seems to support something like the preventative APP. He defends it with a comparison to a scenario involving individual people:

> There are familiar cases where we think that a person is obligated to assist others even when they played no part in the other's poverty or sickness. In such cases we think that a positive duty falls on those able to help. For example, if someone sitting next to you at a table suddenly becomes seriously ill and you're well placed to help, then we tend to think that you should do so. (Caney 2010, 216)

The scope here—the very bad outcome—is "someone's poverty or sickness." One might here already question the analogy to climate change in that "helping" an ailing diner might take little effort, whereas the decarbonization of a state appears to be significantly costly. But assisting a stranger or an acquaintance to prevent a very bad outcome often does come with significant risks or costs. If, for instance, you are helping the ailing person in a somewhat remote forest and emergency services instruct you by phone to wait several hours with the ailing person, we tend to think you should, even if it means you return to the city too late for a crucial job interview. Even if it might turn out to be costly, I agree with Caney's intuition here that I should help remedy my tablemate's illness and I imagine that intuition is widespread, even among the reactionary citizens we are imagining.[16] Presumably, it would be morally unacceptable for the collective at the table to let the diner suffer alone. Those able to assist should assist. It would, after all, be a very bad outcome to let the unfortunate person suffer.

Other authors provide another example of a domestic analogy when they compare the demands of an international climate regime based on the ability to pay to that of a domestic scheme of taxation (Moellendorf 2014, 176, Harris 2019, 2). A supporter of the preventative APP might build this comparison into an argument. To prevent very bad outcomes, a wide range of public goods in our societies—infrastructure, defense, judicial systems, and so on—should be funded via progressive tax systems, which levy the highest costs (and even the highest *rates* of costs) on those most financially able to pay them. The cost of preventing the very bad outcomes of undersupplied public goods, the argument goes, should be divided by ability to pay, and similarly, the burden of preventing very bad outcomes with regard to climate change should be shared according to the same pattern. If we support progressive taxation for projects to protect public goods within a domestic society, the argument goes, it would be inconsistent to deny the underlying principle behind progressive taxation—ability to pay—in the case of protecting the global public good of a safe climate.[17]

Assume, if only for the sake of argument, that the examples of the ailing diner and progressive taxation do support a broad idea that *individual persons* in a *society* should bear the burdens of preventing a very bad outcome in proportion to their ability. This does not yet result in an argument for the preventative APP between states. The move from our intuitions about individuals to a principle that is meant to govern states could progress in two quite different ways, both of which are controversial.

Analogy between Persons and States

The most obvious reading of the arguments above suggests that they are direct analogies. Persons have a duty to prevent very bad outcomes in proportion to their ability to do so. States have duties that mirror those of persons, just at a larger scale. Rich states facing climate costs are then like the tablemates close to the ailing diner or like a wealthy individual considering why she has a duty to pay a higher tax rate.

The political feasibility problem with this direct analogy approach is that reactionary citizens and constrained policymakers will likely be skeptical of the extent to which the moral requirements on states are as stringent as those on persons. Given how different states and persons are (for instance, states do not experience moral emotions), many think it would be strange if the same moral norms applied to both. Even if states and persons were similar enough, it is unclear how much of a community (as opposed to a competition) the collection of states really is (Norman 2020). By contrast, a dinner table or domestic collaboration on a state infrastructure project is not a competitive arena.

Even if the same norms applied in theory, a critic would note that the two actors do not follow moral norms to the same extent. Even putting aside hardnosed realism, in which to talk of states' moral duties is a kind of category mistake (Feaver et al. 2000), few international relations theorists hold that the moral requirements that states might perceive and act upon are simple corollaries of the moral requirements on individuals (Legro and Moravcsik 1999). Rather, a more common position among international relations theorists and political decision-makers is that states act on more minimal norms that align more closely with the "harder" aspects of international law: to keep good faith agreements, obey *jus cogens* norms such as the prohibition of slavery and piracy, to respect the basic human rights of the people in their territory, and to respect the right to self-determination of other states. Given that states and persons behave differently, the political opponent might claim, shouldn't the practical moral requirements on states also be more lenient? It would be little wonder, then, if large proportions of the public in some of the wealthy states to which the preventative APP would attach duties should object that states owe each other less than individuals in the same society do. The direct analogy between the domestic cases and state duties threatens to be unconvincing to key political opponents and thus quite politically infeasible.

State Duties Constructed from Individual Duties

There is another way to try to understand the arguments above that begin with intuitions about domestic situations and make conclusions about international affairs. Rather than directly comparing states to moral agents such as persons, one might see them as emphasizing the duties that individuals have to each other (regardless of state membership) then implying that the moral principles that govern states are a function of those duties. But this strategy for justifying the preventative APP to a reactionary citizen or a constrained policymaker faces a challenge. It requires us to move from interpersonal relationships between compatriots or those immediately around one to relationships that cross state borders. While a broadly cosmopolitan outlook acknowledges no morally relevant difference between the diner one sits next to and the similarly needy individual in another state, a non-cosmopolitan outlook treats state boundaries as morally relevant. If state duties are a function of individual duties and the duties on individuals across state borders are highly disputed, the state duties we can supposedly construct from the duties of individuals across state borders are going to inherit that controversy. And there is a widespread belief that one can choose to favor one's compatriots in terms of aid which I will explore further below. This controversy means that arguments based on constructing duties between states from those that exist between individuals will not help the feasibility of the preventative APP.

Furthermore, even if we accept that individuals have strong duties of charity to needy individuals in other states, this does not get the preventative APP off the ground. The idea that the moral principles that govern states can be constructed from the duties of its members raises further difficulties. It will typically be unclear, at least to the reactionary citizen, or even the constrained policymaker, how such a construction occurs. By what function does the state inherit the duties of its members? What happens to the duties of the individuals when the state assumes them? At the very least, the political opponent could ask for a clear account of the construction of state duties from individual duties. This is likely to look different depending on whether we are methodological individualists who see a state as an institution made up of a collection of individual agents (albeit interacting in special roles) or those who see it as an agent itself, with very different duties to individuals or other states (List et al. 2011).

Public Attitudes toward Development Aid

Regardless of the way the argument from individual duties is meant to proceed (whether by analogy or construction) the kinds of duties of assistance to poorer countries that large parts of the general public in wealthy countries accept seems to belie easy comparison to the duty of assistance between compatriots. Some countries are ailing, just like the sick diner, but assistance from those able to help remains very low. The exact support for development aid among the public is notoriously hard to test because people are typically confused about the facts and try to please interviewers by professing at least a "normal" amount of support for development assistance (Hudson and Van Heerde 2012). However, one U.S. study about financing the Sustainable Development Goals found that 41 percent of respondents (and 64 percent of Republicans) rated "very convincing" an argument that stated "Spending tens of billions to help other countries when we are facing our own severe problems is irresponsible. Each country should focus on solving their own problems and encourage others to do the same" (Program for Public Consultation 2020). This was compared to 25 percent (7 percent of Republicans) who rated the contrary argument in favor of development assistance "very convincing" (Program for Public Consultation 2020). In several European countries, Alain Noel and Jean-Phillipe Therein (2016) analyzed Eurobarometer data to assess the relationship between public concern for domestic inequality and support for development aid. Perhaps surprisingly, they found that the more citizens felt there were serious distributional concerns at home, the less support they gave for development assistance. The likely explanation is a public perception of priority. To the extent that inequality is a significant problem among compatriots, even urgent demands of those elsewhere are treated as less significant (Noel and Therein 2016). Furthermore, the relative lack of prominence of international assistance in national policy debates is evidence for the view that many people place only small weight on any duty to avoid very bad outcomes (if the bad outcomes are felt elsewhere). So understood, the analogy from domestic situations between individuals doesn't seem to make the preventative APP particularly political feasible.

Concern for the Future Poor

Another possible defense of the preventative APP does not try to begin with a principle operative between individuals, but instead looks at the details of the climate change case in particular. For instance, Caney first asks us to consider, in roughest terms, the options for bearing the climate change remainder. Either the current poor bear the burden, the current rich bear the burden, or neither do and the remaining burdens of climate change fall where they may, which is to say, mainly on the future poor. Caney notes that the future poor seem particularly undeserving of extra burdens, and since the rich can bear the burden "without sacrificing any reasonable interests" they are thus the ones who should (Caney 2010, 214).

Caney's argument from intuitions about the climate case relies on somewhat more specific, but no less contested grounds than treating interstate morality as interpersonal morality, or constructing state duties from individual duties. Caney suggests that there is strong reason not to allow both rich and poor countries to ignore climate change caused by the remainder and leave future people to foot the bill. In fact, he sees that the kind of intergenerational buck-passing as exactly the kind of bad outcome that triggers the APP. But this just leads straight to the debate about whether the principle is triggered at all—whether the outcome is bad enough. Philosophical orthodoxy suggests that the interests of future people matter a great deal, if not equally to those alive today.[18] But how to weigh harm to future people is yet another deeply contested topic in the political realm. One could note the policy debates about pure time discounting,[19] or the scarcity of robust institutions to protect future people, or a relative lack of interest in existential risks (Ord 2020) as a sign of public lack of concern about the welfare of future people. And even if there was widespread public agreement that harm to future people from climate change counts as a very bad outcome, some popular arguments suggest that continually rising economic growth means that future people as a group will have resources to adapt to future climate damages. Of course, numerous philosophers have argued for the necessity of taking climate action to prevent harm to future people: to avoid a tragic tyranny (Gardiner 2011), human rights violations in the future (Caney 2009) or because of the irrelevance of continually rising global gross domestic product (GDP) (Jamieson 2014). But the point here is not to find a principle that will convince other philosophers, but to propose a principle that is likely to be marginally persuasive to reactionary citizens and constrained policymakers.

My overall assessment of the preventative APP is that it is not very politically feasible. It places costs on people who are at least tempted by non-cosmopolitan and realist views, but it is very hard to justify without relying on cosmopolitan assumptions or assuming a human-like morality of states. We next turn to a formulation of the APP that is specifically designed to be more politically feasible—the voluntarist APP.

III. VOLUNTARIST APP

The preventative APP has a rather permissive scope: it deals with the distribution of the burdens of preventing some very bad outcome. If we narrow the scope of the APP, can we ameliorate some of the feasibility worries? Consider a voluntarist APP.

> **Ability to Pay Principle (Voluntarist):** The collective burden of meeting *an agreed-upon collective goal* should (where no better principle applies) be borne in relation to a state's ability to pay.[20]

The voluntarist APP mirrors the preventative APP by ceding lexical priority to unspecified "better principles." I simply assume here that a PPP can be formulated in a way that has lexical priority and is equally politically feasible (likely by basing it on the more robust notion of nonaggression between states). At least it will apply specifically to the remainder. But unlike the preventative APP, the voluntarist APP has a restrictive scope. It only applies to the burdens of meeting an agreed-upon collective goal.

One politically feasible justification for the voluntarist APP is this. States owe each other little under normal circumstances. But where they share an agreed-upon collective goal, and particularly where the shared goal has been formalized in an international treaty, this creates a tighter community of agents (with regard to that issue), and the domain becomes more cooperative; thus, considerations of fair burden-sharing become prominent. And where no other obvious division of burdens to meet this shared goal seem relevant, ability to pay is an apt principle by which to divide the burden.

Why should a shared goal create stronger connections between states? One way to consider it is from the game-theoretic perspective. It is in all states' interest to keep global temperatures relatively low. But individual states all have a reason (some more than others) to free-ride and rely on other states to be the suckers that pay to reach the goal. In the absence of a global leviathan, *some* conception of fair burden-sharing is necessary to prevent universal defection in this case, by disgruntled leaders or populaces not wanting to be a sucker. Furthermore, all countries have some reason to prefer an orderly, rules-based international regime, and respecting fair sharing to meet collective goals is a part of that rules-based package. Still, given its low lexical priority, it is not the main or only potential principle for dividing the burdens. I argue, however, that it is a good default principle of division where no better principle applies.[21] Why does it make a good default? Perhaps this is where explanation comes to an end (Shue 1999). However, as we see in the paragraph below, the choice of a voluntarist APP in the past is one reason to continue using it to avoid a collective action problem persisting in the case of climate change. It also avoids the controversial assumptions required by the preventative APP. Unlike the various arguments for the preventative APP, it does not treat states as subject to the same moral constraints as persons, nor does it assume that state duties can be constructed out of individual cosmopolitan duties to aid. Its normative basis—treating states as permitted to pursue their own self-interest, mainly constrained by their own agreements and perhaps requirements of rationality—is much more palatable to many political opponents. Thus, the voluntarist version beckons as a more politically feasible formulation of the APP.

There is a complementary argument for the voluntarist APP. Many political opponents of a conservative bent place importance on the existence of clear precedents in the international domain, and the voluntarist APP does seem to have such precedents. One example is the contributions to the general budget of the United Nations (UN). The commitment to form a cooperative body for peace, security, and prosperity postwar predated the exact sharing of the operating budget. Once general cooperative goals were agreed upon, something like the voluntarist APP was enacted. The operating budget of the UN (around 12 billion USD annually) and dues are linked to states GNI except with a ceiling of 0.01 percent of the budget for the least developed states and an overall ceiling of 22 percent of the budget, which effectively only applies to the United States.[22] Likewise, when the Pearson Commission convened during the late 1960s to discuss how developed countries should divide the burden of the shared goal of promoting global development, a proportion of countries' GNP (eventually 0.7 percent) was chosen as the relevant sharing rule. The UN General Assembly adopted this rule by consensus as part of the International Development Strategy in 1970. Since then, the 0.7 percent of GNP rule among developed countries has been affirmed many times.[23] While neither the general budget contributions rule nor the Pearson Commission 0.7 percent goal is being complied with completely, these examples provide evidence that a voluntarist APP is already part of the state-based international system. This increases the political feasibility of the principle to constrained actors. Such actors can suggest to those who constrain them that while the scale of the challenge of climate change is new, the principle of division is not. Reactionary citizens might still dismiss these examples as more cases of problematic globalism. However, others might accept these precedents, especially if they see international law as a slow accretion of techniques to solve collective action problems among self-interested actors.

Applying the Voluntarist APP

We have justified the voluntarist APP and suggested it is more politically feasible than the preventative APP. Does the voluntarist APP apply to the problem of climate change? Yes, it does. First, the scope is triggered. The clearest aspects of international climate law are perhaps now the shared goals. "The ultimate objective" of the collection of the 197 states that have ratified the UNFCCC is "to achieve . . . stabilization of greenhouse gas concentrations in the atmosphere at a level that would prevent dangerous anthropogenic interference with the climate system" (UNFCCC 1992, art. 2). The Paris Agreement gets more specific, affirming that Parties collectively "aim to strengthen the global response to the threat of climate change . . . by . . . holding the increase in the global average temperature . . . to well below 2°C . . . increasing the ability to adapt to the adverse impacts of climate change . . . and making finance flows consistent with a pathway toward low greenhouse gas emissions and climate-resilient development (UNFCCC 2015a, 2.1).[24] It seems fair to say that states have a set of shared climate goals, which creates a community for this issue that should be governed by fair sharing rules.

Second, although the voluntarist APP, like the preventative APP, has low lexical priority, no better principle applies to the remainder as a practical principle for fair division. A PPP is a nonstarter, since we are discussing the distribution of the remainder, which by definition cannot be distributed via a PPP. What about a beneficiary pays principle, by which those who have benefited from past emissions should bear the costs associated with them? This has several problems. Part of the remainder is the emissions from those too poor to be asked to bear burdens, but it is primarily the poorer countries that benefit most from the emitting activity of the global poor. Thus, under a beneficiary pays principle, that part of the remainder would either have to be borne by the poorer countries (not particularly politically feasible either) or not borne at all. The other part is the burden from historical emissions. This burden could be passed on to wealthy countries under a beneficiary pays principle, but that move is not politically feasible. The idea that the beneficiaries of decisions made generations ago should pay for the consequences of those decisions is likely to be seen by political opponents in rich countries as a reductio ad absurdum of a historical beneficiary pays principle that reaches too far into the past and blames the innocents for what they could not have known was harmful. Finally, even if the beneficiary pays principle were to be a good candidate in theory, it would be particularly difficult to apply in practice.

What about a principle that is more lenient on wealthy countries? Why not propose the hyperrealist version of the beneficiary pays principle: that those who benefit most from the global *effort to solve climate change* should bear the greatest burden (Posner and Weisbach 2010)? The United States would be more willing to move on climate if it received side payments from Bangladesh: everyone wins! Such an arrangement would be stable too, if the most vulnerable states can put aside their pride· But even if it would be accepted by the most vulnerable, as Stephen Gardiner (2016) points out, this would be essentially for the wealthy countries to run an extortion racket. Treating fair sharing of an agreed-upon goal as an extortion racket treats the project to respond to climate change as a competitive arena rather than a cooperative zone.[25] This in turn would mean that those who entered the Framework Convention and the Paris Agreement acted in bad faith.

No other politically feasible principle applies well to the remainder. Thus, unless we can uncover a better principle, APP wins by default as a focal point for distribution of an agreed-upon burden.

The Voluntarist APP in International Climate Law

At least under a popular interpretation, international climate law also affirms the APP is one of the principles in play. The 1992 Convention states, broadly, that "parties should protect the climate system . . . in accordance with their common but differentiated responsibilities and *respective capabilities*" (CBDR-RC), where capabilities were widely understood to be the stark divide in wealth and income between two worlds—the developed and developing (UNFCCC 1992, art. 3.1).[26] The Paris Agreement is, in the words of the Durban Platform, "under the Convention." It is guided to a large extent by its principles and refers to the Convention forty-eight times (UNFCCC 2011, 2015a). The Paris Agreement also makes specific reference to, and updates, the language of CBDR-RC. Paris Agreement Article 2.2. states: "This Agreement will be implemented to reflect equity and the principle of common but differentiated responsibilities and respective capabilities, in the light of different national circumstances."[27] Furthermore, the Paris Agreement consistently references an expectation that developed states "should continue taking the lead" on mitigation by undertaking economy-wide absolute emission reduction targets" while "developing state Parties . . . should continue enhancing their mitigation efforts" (ibid. art 4.4). Arguably, both the principles of equity in the UNFCCC and Paris Agreement, and their overall structure follow the voluntarist APP: a collective goal is named, and both contribution (perhaps with lexical priority), and ability to bear burdens to address the problem are noted as relevant for meeting the goal. But the argument for the voluntarist APP in this context is not simply that a reference to ability to pay happened to be negotiated into place in international climate law and that states should keep their agreements.[28] Rather, even under conservative views of international duties, once the prospect of a cooperative, shared goal is in play, a reference to ability to pay is an *appropriate* default principle to guide division of that goal that is mirrored in international climate law.

To sum up, restricting the scope of the APP to situations where actors have antecedently committed to contribute to a shared goal makes it considerably more feasible. States have already agreed to a shared collective goal: preventing dangerous interference in the climate system and keeping the global temperature rise "well below" 2°C. After more obvious principles like a PPP have been applied, using a voluntarist APP to distribute the remainder of that burden can be argued for more persuasively to those likely to be starting from more conservative, statist intuitions. Where a preventative APP draws on broadly cosmopolitan intuitions about how states' duties to bear the climate burden might mirror or reduce to individuals' duties, a voluntarist APP can be defended to political opponents as a way to solve tough collective action problems among a group at least nominally committed to the cooperative goal. Furthermore, this division has an agreed-upon collective goal that is affirmed in the UNFCCC and has echoes in other shared projects. The voluntarist APP holds significant promise as a politically feasible burden-sharing principle in the case of climate change. In the next section, I turn to objections to the voluntarist APP.

IV. OBJECTIONS

Objection from Ephemerality

Here I raise an objection about the voluntarist APP: It is too ephemeral as a basis for climate justice. If the only trigger for the duty to bear the climate change remainder were these voluntary agreements made by states, this entails that any wealthy state that wanted to shirk their moral requirement to bear the climate change remainder could choose to lessen their burden by simply withdrawing from the relevant agreement, and thus the collective goal. The critic fears that the voluntarist APP thus sets out not a principle of climate justice, but a contingent agreement that can be undermined at the whim of powerful states.

As a practical matter, the objection from ephemerality does not seem particularly salient. While the United States stated its withdrawal from the Paris Agreement on June 1, 2017, no other state has followed suit, and the United States is likely to have rejoined by the time this book is published. After a twenty-three-year struggle to achieve a new comprehensive global treaty on climate change, ratified by an astonishing 187 states, it is not surprising that many other states will be eager to try to renegotiate the key principles of the

Paris Agreement. This is compounded by the highly conservative nature of the UNFCCC's decision-making process, which relies on consensus, not voting, to make any change. As Moellendorf points out, citing Robert Keohane, international agreements are much harder to come by than to maintain, and it is hard to imagine states giving up the hard-fought compromises of Paris to start from scratch or go it alone (Moellendorf 2016). It is also interesting that despite President Trump's dismissive attitude toward the Paris Agreement, there has been no signal that the United States considered pulling out of the 1992 UNFCCC.[29] It is the UNFCCC that provides the overarching (and demanding) goals of preventing "dangerous anthropocentric interference" in the climate system, and reiterates the principle of meeting that goal in light of states' "common but differentiated responsibilities and respective capacities." The ephemerality of international agreements should not be overstated.

Objection from Irrelevance

There is a second kind of objection to the voluntarist APP as the salient principle for political feasibility. The objection is that it is not politically feasible after all, for key constituencies (reactionary citizens and constrained policymakers from rich countries) simply do not accept that international law is "real law" and thus is not of any particular moral relevance. States can sign agreements, the argument goes, but still act roughly as they wish, since such treaties contain minimal, if any, enforcement. In fact, the objection goes, they decide very often to bend or violate treaties they have ratified. Consequently, their citizens and bureaucrats feel no great compulsion to respect the force of agreed-upon collective goals in climate treaties.

This objection underestimates the power of international law. While international law has never been, and arguably should never be, enforced like the law of a sovereign state is, this does not mean it is toothless and irrelevant. In fact, much work in international studies has tried to explain why states obey international law so often *despite* the lack of direct enforcement. In an oft-quoted line Louis Henkin, former president of the American Society of International law, notes that "almost all nations observe almost all principles of international Law and almost all of their [legal] obligations almost all of the time" (Henkin 1979). As one of the major international law textbooks puts it, we are hostage somewhat to availability bias: violations of international law are newsworthy, while most states following most of the thousands of international legal instruments is not (Shaw 2008).[30] Most scholars are particularly familiar with the conduct of the United States, a country with a difficult relationship with international law, to put it mildly. But even the United States rarely violates international law. Rather, it refuses to sign, refuses to ratify, or argues for special "exceptions and reservations" written into law before signing and ratifying. A final consideration is the spectrum of international law, from hard law (treaties, jus cogens laws against piracy and slavery) to soft law (political declarations and UN General Assembly resolutions). For instance, it is more common for countries to act contrary to General Assembly resolutions, but the majoritarian nature of most of those decrees gives them much less gravitas than treaties and agreements such as the Framework Convention and the Paris Agreement, negotiated as they were through painstaking consensus processes.

As far as political feasibility goes, the widespread actual compliance with international law is enough to save the voluntarist APP as a justification to constrained policymakers for rich countries bearing the climate remainder. But what about reactionary citizens? Would they care whether or not their state has joined a cooperative venture to prevent dangerous interference with the earth's climate system? Again, the falling fortunes of the international order in recent years might threaten the idea that reactionary citizens, especially, put any stock in something like the ability of international law to create cooperative communities on an issue. This is a concern, although it is too early to say what effect the global pandemic or the rise of China as a superpower will have on public respect for the international order. Still, we have some recent social science evidence that individual citizens, including conservatives, still pay some heed to international law. A survey experiment of several thousand subjects in India, Australia, and the United States described a refugee policy supported by the country's leader, then informed some subjects of the international illegality of the policy, while informing others of the leader's support for the policy. They found that the international law information reduced subjects' support for the policy by a "small but significant" amount (Strezhnev, Simmons, and Kim Forthcoming). Even reactionary citizens might take the commitment to collective climate goals seriously enough to make the voluntarist APP feasible.

Worry: Loss and Damage

I will consider here a worry. The voluntarist APP supported by a persuasive legal argument has too narrow a scope. The voluntarist APP applies to agreed-upon goals, which we saw above were the collective burdens of mitigation and adaptation, and providing climate finance to assist these goals. But one kind of burden that would not necessarily trigger a voluntarist APP is the burden of compensating for "loss and damage." Loss and damage, in international negotiations, refer to the harms of climate change that will inevitably occur despite efforts at adaptation or mitigation. Compensation for loss and damage does not easily fall under the category of "agreed-upon collective goal." After all, the Framework Convention refers to "preventing" not "repairing" dangerous interference. Article 8 of the Paris Agreement only contains the mild language that states, "recognise the importance of averting, minimizing and addressing loss and damage" (UNFCCC 2015a, art. 8.1), without any acknowledgment that this is a shared goal or specific duty. The accompanying Paris Decision, furthermore, (in)famously states that parties agree "that article 8 does not involve or provide a basis for any liability or compensation" (UNFCCC 2015b: para. 51).

At this point, a supporter of the voluntarist APP should probably bite the bullet. Under current conceptions of the shared goal of responding to climate change, the voluntarist APP does not require wealthy states to pay compensation for damage that occurs despite mitigation and adaptation efforts. If we presume that there is a clear moral case for wealthy states paying loss and damage, the voluntarist APP might be seen to trade off climate justice for feasibility. But it is important not to overstate this result. After all, the voluntarist APP does not rule out arguments that highly polluting states should compensate for loss and damage *caused by their pollution*. The usual caveats about emissions produced under ignorance and making the PPP poverty-sensitive might apply, which could presumably limit the proportion of loss and damage that would require compensation. This means that the voluntarist APP in combination with a PPP leaves open the possibility for a principled compromise on loss and damage. That part of loss and damage that was caused by the remainder falls where it may and that part captured by an appropriately bounded PPP falls to the polluter. Loss and damage caused by the remainder might be a very bad outcome, but it does not fall under the agreed-upon collective goal (yet), and requiring it to be borne in relation to ability to pay is not currently politically feasible. (For more on distributing the burden for loss and damage, please see Hossain et al. in this volume.)

V. CONCLUSION

I have laid out an argument for bearing the climate change remainder with a specific, voluntarist APP. It is a formulation of the APP that is more easily justified to political opponents than typical formulations. Because it is more easily justifiable to such opponents, it makes the world it describes more politically feasible, more likely to come to pass in an environment marked at times by bitter division and parochial worldviews. Of course, I have said nothing about how to measure ability to pay, and it is likely to be in an operational stage that more bitter disputes arise, particularly between industrialized states and rapidly emerging economies.[31] But my main point is there is no need to saddle the APP at the outset with a scope that is so contentious that it could not be justified to constrained policymakers or reactionary citizens.

Some might fear that the justification for my voluntarist APP rests on a flimsy foundation of vague treaties only held together by the consent of states. I have tried to allay those fears. However, I admit that my specific APP will not apply to the burden of loss and damage. This might be enough to lead those committed to progressive goals or ideal theory to reject a voluntarist formulation of an APP. To those I ask only for an account of what philosophizing about climate justice is for. If one thinks it is only to witness our moral failure, utopian theories might suffice. But what if a legitimate purpose of climate philosophy is to prepare policymakers to justify going beyond a state's narrow self-interest or encourage them to engage with reactionary citizens? Or to attempt to persuade reactionary citizens directly? In that case, I suspect this minor weakening of the APP will be repaid many times over by its greatly increased feasibility. (For more on the role of philosophers in climate change policymaking, please see Kowarsch & Lenzi in this volume.)

NOTES

1. This example is from Lawford-Smith (2013).
2. Although these are sometimes relevant factors for the difficulty of political transformation; see also (Brennan and Sayre-McCord 2016).
3. (Carens 2013) See (Southwood 2019) for an argument that this framing is misleading.
4. See Stephen Gardiner: "Though one purpose of ethics is to guide action, in my view it also plays a role in *bearing witness* to serious wrongs even when there is little chance of change" (Gardiner 2016, 255). See also (Kyle Fruh, n.d.).
5. See, for example, Posner and Weisbach (2010) and Weisbach (2016) for rather extreme versions of these views.
6. See the self-admitted failure of Michael Ignatieff's Canadian political career (Ignatieff and Warburton 2014).
7. In *some* senses, this goal could be exemplified by the first three sections of chapter 3 of the IPCC's Fifth Assessment Report by Working Group III, which translate philosophical positions in a manner and forum to be consumed by policymakers (Working Group III. 2014). However, to the extent that it summarizes work that is not concerned with political feasibility, it does not fit our present purposes well.
8. By "compensate" I refer to the process of paying for the "loss and damage" resulting from climate change despite efforts at mitigation and adaptation.
9. See the IPCC's Special Report on Global Warming of 1.5°C, especially Section 4.21. "Both the integrated pathway literature and the sectoral studies agree on the need for rapid transitions in the production and use of energy across various sectors, to be consistent with limiting global warming to 1.5°C. The pace of these transitions is *particularly significant for the supply mix and electrification*" (IPCC 2018: 320)

10. See (Francis, online ahead of print) for a defense against the objection that treating states as the relevant moral agents creates unfair burdens on individuals.

11. For a general example, see (Miller 2001). For an application to climate change, that is not discussed below, see (Miller 2008), where the scope is the violation of human rights from unmitigated climate change.

12. Some even support an APP that does not cede lexical priority in this way (Moellendorf 2014: ch. 6). For such an account, the politically feasibility of the APP is even more important.

13. See (Höhne, Den Elzen, and Escalante 2014) for a review.

14. As in (Baer et al. 2008) and operationalized by the Civil Society Review (2015). See also the argument in (Shue 2015). Edward Page suggests a role for the APP as a means of separating out those states that lack any burdens at all from those that bear burdens (Page 2008).

15. NDCs are "Nationally Determined Contributions"—states' official mitigation pledges under the Paris Agreement. The websites in question include www.parisequitycheck.org, www.climateactiontracker.org, and www.civilsocietyreview.org.

16. There is disagreement of course, about whether the law should punish those that do not assist in relatively easy rescues, even if they are able to. But that is a different matter (Feinberg 1984).

17. While this domestic analogy is suggested by Moellendorf, he has a different, and in my view, more politically feasible argument for a more voluntarist APP, on the basis of there being a right to sustainable development in the UNFCCC. I disagree that such a right was actually agreed upon in any hard law instrument (see Biniaz 2016), but the form of his argument is promising (Moellendorf 2014, 2016).

18. (Sikora and Barry 1978). For an example of a philosopher (with an economist) bucking the orthodoxy, see Beckerman and Pasek (2001).

19. See the vast literature on discounting. A good introduction is in (Broome 2012).

20. Shue's version of an APP (about environmental problems in general, not climate change in particular) could be seen to be a voluntarist APP: "Among a number of parties, all of whom are bound to contribute to some common endeavor, the parties who have the most resources normally should contribute the most to the endeavor" (Shue 1999, 537). The formulation is deliberately vague about what it means to be bound. Shue provides one justification from broad sufficientarian intuitions, but considers that "the general principle itself is sufficiently fundamental that it is not necessary, and perhaps not possible, to justify it by deriving it from considerations that are more fundamental still" (Shue 1999, 537).

21. Compare Schelling's idea of a focal point—an otherwise arbitrary point chosen for coordination purposes to avoid a social dilemma (Schelling 1980).

22. These figures are market exchange rates, not purchasing power. Admittedly, the application of the burden-sharing scheme for the UN does not have full compliance. A large (and growing) proportion of the costs of the UN are funded by voluntary contributions, which has been flagged as a cause for concern (Yussuf, Larrabure, and Terzi 2007). Notably, besides receiving special consideration in reducing its proportion of the costs (via the 22 percent ceiling), the United States has at times fallen significantly in arrears in its payments.

23. See (OECD 2002). In the DAC Aid Review of 1971, the United States stated that it rejected "specific targets." However, the spirit of the principle was still respected by affirming that states "should make their best efforts" to increase development assistance (OECD, 2002).

24. This strengthened a specific temperature goal first agreed-upon in a UNFCCC Decision at Cancun in 2010 (UNFCCC 2010, I.4).

25. Indeed, it is no longer clear why solving climate change is a shared goal at all, for the side payments to rich countries could be made as a series of discrete transactions.

26. As evidenced by the division of states into Annexes: Annex 1 states were given mitigation commitments, the subset Annex 2 states (excluding "economies in transition") were given both financial and mitigation commitments, while non-Annex states had no substantive mitigation or financial commitments.

27. The addition of "in the light of different national circumstances" reflects a compromise between developed states and emerging economies. It balances the importance of the 1992 division of states into Annex/non-Annex parties on the one hand, and the rising emissions and wealth associated with emerging economies on the other hand (Rajamani 2016).

28. As Darrel Moellendorf argues (in work that has significantly inspired this chapter), taking these norms of the UNFCCC seriously can provide arguments for climate justice that have more bite in the international realm. This is because they rest not on any particular view of global justice, but on the "promissory obligations" that states voluntarily took on by signing or ratifying these treaties (Moellendorf 2016). Indeed, that promissory obligation rests in the oldest principle of international law, *pacta sunt servanda* (agreements must be kept) (Shaw 2008).

29. It might be argued that the Trump administration was simply unaware of the commitments in the UNFCCC Framework Convention, and fundamentally disagreed with the idea of anthropocentric interference on the climate or the desirability to take steps to limit it. But even Donald Trump's own record on climate denial is patchy and opportunistic (Friedman 2020). His withdrawal speech focused on the supposed unfairness of the distribution principles, not climate denial or an attitude that climate change would not be bad.

30. There are roughly 3,700 international *environmental* agreements in the IEA database alone (https://iea.uoregon.edu/).

31. Indeed, the plenary in the negotiations at Paris largely concerned whether the world had "changed" significantly from 1992 in terms of countries ability to pay or not, with the chief negotiator for the EU and the negotiator for the "like-minded-developing-countries" bloc trading statistics about economic development among middle-income countries (author's notes).

REFERENCES

Baer, Paul, Tom Athanasiou, Sivan Kartha, and Eric Kemp-Benedict. 2008. *The Greenhouse Development Rights Framework: The Right to Development in a Climate Constrained World* (Revised 2nd Edition). Berlin: Heinrich Böll Foundation.

Beckerman, W., and J. Pasek. 2001. *Justice, Posterity, and the Environment.* Oxford: Oxford University Press.

Biniaz, Susan. 2016. "Comma but differentiated responsibilities: Punctuation and 30 other ways negotiators have resolved issues in the international climate change regime." *Michigan Journal of Environmental and Administrative Law* 6(1): 37–63.

Brennan, Geoffrey, and Geoffrey Sayre-McCord. 2016. "Do Normative Facts Matter... To What Is Feasible?" *Social Philosophy & Policy* 33(1–2): 434–456.

Broome, John. 2012. *Climate Matters: Ethics in a Warming World.* New York: WW Norton.

Buchanan, Allen. 2002. "Political Legitimacy and Democracy." *Ethics* 112(4): 689–719.

Caney, Simon. 2009. "Climate Change and the Future: Discounting for Time, Wealth, and Risk." *Journal of Social Philosophy* 40(2): 163–86.

Caney, Simon. 2010. "Climate Change and the Duties of the Advantaged." *Critical Review of International Social and Political Philosophy* 13: 203–228.

Carens, Joseph. 2013. *The Ethics of Immigration.* Oxford: Oxford University Press.

Civil Society Review. 2015. "Fair Shares: A civil society review of NDCs."

Feaver, Peter D, Gunther Hellmann, Randall L Schweller, Jeffrey W Taliaferro, William C Wohlforth, Jeffrey W Legro, and Andrew Moravcsik. 2000. "Brother, Can You Spare a Paradigm? (Or Was Anybody Ever A Realist?)." *International Security* 25: 165–193.

Feinberg, Joel. 1984. *Harm to Others.* Oxford: Oxford University Press.

Francis, Blake. Online ahead of print. "In Defense of National Climate Change Responsibility: A Reply to the Fairness Objection." *Philosophy and Public Affairs.*

Friedman, Lisa. 2020. "With the cameras off, Trump softened his climate denial." *New York Times*, October 22, 2020. https://www.nytimes.com/live/2020/10/22/us/trump-biden-debate-tonight?partner=naver#with-the-cameras-off-trump-softened-his-climate-denial

Fruh, Kyle n.d. "Climate Change Driven Displacement and Anticipatory Moral Failure." *In Preparation.*

Gardiner, Stephen 2011. *A Perfect Moral Storm: The Ethical Tragedy of Climate Change.* Oxford: Oxford University Press.

Gardiner, Stephen. 2016 "In Defense of Climate Ethics." In *Debating Climate Ethics*, edited by Stephen Gardiner and David A. Weisbach, 3–86. Oxford: Oxford University Press.

Gilabert, Pablo, and Holly Lawford-Smith. 2012. "Political Feasibility: A Conceptual Exploration." *Political Studies* 60(4): 809–825.

Harris, Paul. 2019. "Introduction." In *A Research Agenda for Climate Justice,* edited by Paul Harris (1–14). Cheltenham, UK and Northampton, MA, USA: Edward Elgar Publishing.

Henkin, Louis. 1979. *How Nations Behave: Law and Foreign Policy.* New York: Columbia University Press.

Höhne, Niklas, Michel Den Elzen, and Donovan Escalante. 2014. "Regional GHG reduction targets based on effort sharing: a comparison of studies." *Climate Policy* 14: 122–147.

Hudson, David and vanHeerde-Hudson, Jennifer. 2012. "A Mile Wide and An Inch Deep: Surveys of Public Attitudes Towards Development Aid." *International Journal of Development Education and Global Learning* 4(1): 5–23.

Ignatieff, Michael, and Nigel Warburton. 2014. "Michael Ignatieff on Political Theory and Political Practice." *Philosophy Bites Podcast*, April 12, 2014. https://philosophybites.com/2014/04/michael-ignatieff-on-political-theory-and-political-practice.html.

IPCC. 2018. *Global Warming of 1.5°C. An IPCC Special Report on the impacts of global warming of 1.5°C above pre-industrial levels and related global greenhouse gas emission pathways, in the context of strengthening the global response to the threat of climate change, sustainable development, and efforts to eradicate poverty*, edited by V. Masson-Delmotte, P. Zhai, H.-O. Pörtner, D. Roberts, J. Skea, P.R. Shukla, A. Pirani, W.

Jamieson, Dale. 2014. *Reason in a Dark Time: Why the Struggle against Climate Change Failed--and What It Means for Our Future*. Oxford: Oxford University Press.

Kingston, Ewan. 2014. "Climate Justice and Temporally Remote Emissions." *Social Theory and Practice* 40: 281–303.

Kingston, Ewan. 2020. "Assessing the Assessments of NDC Fairness Comparisons." Presented at *Bridging Gaps of Affluence, Nation, and Time*. CFI online workshop.

Lawford-Smith, Holly. 2013. "Understanding Political Feasibility." *Journal of Political Philosophy* 21(3): 243–259.

Legro, Jeffrey W., and Andrew Moravcsik. 1999. "Is anybody still a realist?" *International Security* 24: 5–55.

List, Christian, and Philip Pettit. 2011. *Group Agency: The Possibility, Design, and Status of Corporate Agents*. Oxford: Oxford University Press.

Miller, David. 2001. "Distributing Responsibilities." *Journal of Political Philosophy* 9(4): 453–71.

Miller, David. 2008. "Global Justice and Climate Change: How Should Responsibilities be Distributed?" In *The Tanner Lectures on Human Values*. Tsinghua University, Beijing March 24–25, 2008.

Mittiga, Ross. 2019. "Allocating the Burdens of Climate Action: Consumption-Based Carbon Accounting and the Polluter-Pays Principle." In *Transformative Climates and Accountable Governance*, edited by Beth Edmondson and Stuart Levy, 157–194. Switzerland: Palgrave Macmillan.

Moellendorf, Darrel. 2014. *The Moral Challenge of Dangerous Climate Change: Values, Poverty, and Policy*. Cambridge: Cambridge University Press.

Moellendorf. 2016. "Taking UNFCCC Norms Seriously." In *Climate Justice in a Non-Ideal World*, edited by Clare Heyward and Dominic Roser, 104–121. Oxford: Oxford University Press.

Noel, Alain, and Jean-Phillipe Therein. 2016. "Public Opinion and Global Justice." *Comparative Political Studies* 35(6): 631–656.

Norman, Wayne. 2020. "The Ethical Adversary: How to Play Fair When You Are Playing to Win." Unpublished Manuscript.

OECD (Organisation for Economic Co-operation and Development). 2002. "History of the 0.7% ODA Target." *DAC Journal* 3(4): III-9–III-11.

Ord, Toby. 2020. *The Precipice: Existential Risk and the Future of Humanity*. New York: Hachette Books.

Page, Edward. 2008. "Distributing the Burdens of Climate Change." *Environmental Politics* 17: 556–575.

Posner, Eric and David Weisbach. 2010. *Climate Change Justice*. Princeton: Princeton University Press.

Program for Public Consultation. 2020. "Americans on US Contribution to Five Sustainable Development Goals." http://www.publicconsultation.org/wp-content/uploads/2020/07/SDG_Report_0720.pdf.

Räikkä, Juha. 1998. "The Feasibility Condition in Political Theory." *Journal of Political Philosophy* 6(1): 27–40.

Rajamani, Lavanya. 2016. "Ambition and Differentiation in the 2015 Paris Agreement: Interpretative Possibilities and Underlying Politics." *International and Comparative Law Quarterly* 65: 493–514.

Rawls, John. 1999. *A Theory of Justice*. Cambridge: Belknap Press.

Schelling, Thomas C. 1980. *The Strategy of Conflict*. Cambridge MA: Harvard University Press.

Shaw, Malcolm N. 2008. *International Law*. Cambridge: Cambridge University Press.

Shue, Henry 1993. "Subsistence Emissions and Luxury Emissions." *Law & Policy* 15: 39–59.

Shue, Henry. 1999. "Global Environment and International Inequality." *International Affairs* 75: 531–545.

Sikora, Richard I., and Brian M. Barry. 1978. *Obligations to Future Generations*. Philadelphia: Temple University Press.

Southwood, Nicholas. 2018. "The Feasibility Issue." *Philosophy Compass* 13(8): 1–13.

Southwood, Nicholas. 2019. "Feasibility as a Constraint on 'Ought All-Things-Considered', But Not on 'Ought as a Matter of Justice'?" *The Philosophical Quarterly* 69(276): 598–616.

Strezhnev, Anton, Beth A. Simmons, and Matthew D. Kim. 2019. "Rulers or Rules? International Law, Elite Cues and Public Opinion." *European Journal of International Law* 30(4): 1281–1302.

UNFCCC (United Nations Framework Convention on Climate Change).1992. United Nations Framework Convention on Climate Change. FCCC/INFORMAL/84.

UNFCC (United Nations Framework Convention on Climate Change). 2010. "The Cancun Agreements: Outcome of the work of the Ad Hoc Working Group on Long-term Cooperative Action under the Convention." FCCC/CP/2010/7/Add.1 https://unfccc.int/documents/6527

UNFCC (United Nations Framework Convention on Climate Change). 2011. "Establishment of an Ad Hoc Working Group on the Durban Platform for Enhanced Action." FCCC/CP/2011/L.10 https://unfccc.int/documents/7089

UNFCC (United Nations Framework Convention on Climate Change). 2015a. "Paris Agreement." FCCC/CP/2015/L.9/Rev.1 https://unfccc.int/documents/9064

UNFCC (United Nations Framework Convention on Climate Change). 2015b. "Paris Decision Text." FCCC/CP/2015/10/Add.1 https://unfccc.int/documents/9097

Weisbach, David. 2016. "The Problem with Climate Ethics." In *Debating Climate Ethics*, edited by David Weisbach and Stephen Gardiner, 137–244. Oxford: Oxford University Press.

Wiens, David. 2015. "Political Ideals and the Feasibility Frontier." *Economics & Philosophy* 31 3): 447–477.

Wiens, David. 2016. "Motivational Limitations on the Demands of Justice." *European Journal of Political Theory* 15(3): 333–352.

Wollner, Gabriel. 2013. 'The Third Wave of Theorizing Global Justice: A Review Essay." *Global Justice: Theory Practice Rhetoric* 6: 21–39.

Working Group III to the Fifth Assessment Report of the Intergovernmental Panel on Climate Change. 2014. *Climate Change 2014: Mitigation of Climate Change*. Cambridge: Cambridge University Press.

Yussuf, Muhammad, Juan Luis Larrabure, and Cihan Terzi. 2007. "Voluntary Contributions in United Nations System Organizations: Impact on programme delivery and resource mobilization strategies." Geneva: UN Joint Inspection Unit.

Chapter 5

Climate Justice, Inherited Benefits, and Status Quo Expectations

Lukas H. Meyer

In order to limit the global temperature rise to a maximum of 1.5° Celsius, global emissions must be substantially reduced (see The Paris Agreement in force since November 4, 2016 (United Nations Framework Convention on Climate Change [UNFCC] 2015)). Under this requirement, emissions (or emission rights) are a limited resource. The remaining global carbon or emissions budget has to be shared fairly between states.[1] Given the positive correlation between well-being and emissions levels, such a distribution will have a significant impact on states' options for action and the well-being levels that can be achieved within states in the future. The fair allocation of the remaining global carbon budget (GCB) is therefore of great importance.

From a purely normative point of view, several factors have to be taken into account in this allocation between states. One potentially important, yet controversial, factor is states' past or historical emissions. Should they be taken into account when determining the distribution of the remaining GCB and, if so, how exactly (see Meyer and Sanklecha 2017b, 1–21)?

There are two potential positions on this issue. First, some call for the high emitters' shares of the remaining global budget to be lower due to the emissions that have already taken place within their borders. This view is often put forward by countries with historically and to-date low-emission levels, who therefore claim a larger proportion of the remaining budget for themselves. I discuss arguments that support this position in the section "The Normative Significance of Historically Shaped Legitimate Expectations." A second position is often taken by states with emissions levels that are well above average (in short: high emitters). They take the position that due to the high level they have reached today and have come to rely on, they are entitled to a large share of the remaining emissions budget. This view can be referred to as "grandfathering" emissions rights. National self-interest demonstrably determines various states' negotiating positions on this issue. For example, while highly industrialized countries do not explicitly argue for grandfathering, their nationally determined contributions (NDCs) under the Paris Agreement can be shown to be based on that notion in effect (for empirical evidence, see Williges et al. [under review, Supplementary Information Text]). I discuss what I take to be the most plausible reason that supports this position below.

According to both positions, historical emissions are normatively significant. And importantly, both views seem justifiable as there are developed arguments for both (see Meyer and Roser 2010; Meyer 2013; Meyer and Sanklecha 2011; 2014). In this chapter, I aim to show to what extent the two positions described above are theoretically linked because the same normative reasons for considering past emissions speak for both positions. This seems to suggest that both positions are relevant for the allocation of the remaining emissions budget, but each in a contrary direction that either reduces or increases the fair proportion of high-emitting states.[2] I will argue, however, that this is not correct. This is because it turns out that the normative implications of the two positions are different. I contend that the first argument (the argument of inherited net benefits) is normatively relevant for the allocation of the global budget among states. It speaks in favor of the reduction in the share of (historically) high-emitting states. The second argument (the argument of legitimate expectations) is normatively relevant for the fair national distribution of the burdens of transformation to low-carbon. It does not support an increase in the share of (historically) high-emitting states. Rather, the particularly high costs imposed on individuals in the transformation process that result from the frustration of their legitimate status quo-expectations are to be taken into account in the choice of state strategies for transformation and to be evened out if that is what domestic justice requires. In the end, I argue that my view on the normative significance of the unequal consequences of historical and past emissions for the fair allocation of the remaining global budget of emissions is not only justified but also feasible, at least in a dynamic sense. While an agreement that allocates the global emissions budget in a way that is adequately sensitive to historic emissions is not one that would be feasible for high-emitting states today, all states ought to commit themselves to change the feasibility constraints in such ways that implementing a justly historically sensitive allocation will become feasible for them.

I. THE NORMATIVE SIGNIFICANCE OF HAVING RECEIVED BENEFITS FROM EMISSIONS-GENERATING ACTIVITIES

Before I discuss the significance of historical emissions for the fair allocation of the remaining global emissions budget, I need to clarify why we have reason to care about a fair distribution of the remaining permissible emissions in the first place. Nowadays, almost all of our activities have emissions as a (so far) unavoidable by-product, including the consumption and production of industrial goods, agriculture, and personal and commercial transport. Causing emissions has been a *conditio sine qua non* of most actions that contribute to our well-being. While we have no reason to be interested in emissions as such, we have strong reasons to be concerned about our well-being and thus in being allowed to cause emissions or consume goods so long as emissions are an unavoidable side-effect of the production, trade, and consumption of goods. In other words, what counts are not emissions as such; rather, what counts are the benefits that people realize in carrying out actions that unavoidably have emissions as a side product. Accordingly, I propose to understand "distributing emissions" as the shorthand for distributing the permissions to perform activities that regularly benefit those who engage in these activities and that have emissions as their direct or indirect side-effect. These permissions are often referred to as "emission rights." Distributing emissions therefore stands for distributing the benefits of engaging in activities that directly or indirectly cause emissions by distributing emission rights.

In distributing the GCB for the transformation process to net zero, we should be concerned about distributing the possible benefits that can be achieved through emissions-generating activities. For the transformation period, there is no strong reason to believe that states could benefit more or less from emissions that have been allocated to them, that is, if we assume the global tradability of emission rights. At the same time, the larger a state's share of GCB the less costly this state's transformation to net zero is likely to be or the more benefits the state is likely to be able to realize in the transformation period. As will become clear, my view is that the limited GCB should be allocated on the basis of an equal per capita (EPC) approach and that this distribution should be historically sensitive: the fair allocation of the budget should reflect the benefits that people/states have realized and will realize from past emission-generating activities. That means, I will argue that some parts of historical and past emissions are significant in so far as they are the side product of past activities that have unequal beneficial consequences for currently living and future people.[3]

Let us first observe that the welfare level of a country or region correlates strongly and positively with the historical and current emission levels of these countries and regions. Highly industrialized countries are causally responsible for over three times as many emissions between 1850 and 2002 as developing countries (cf. Baumert et al. 2005).[4] The industrialization of today's high-income countries has caused a very large proportion of greenhouse gases in the atmosphere. The majority of people who live and will live in developing or emerging countries not only have fewer benefits from emissions-generating activities, but they will suffer disproportionately more as a result of the consequences of climate change (cf. United Nations Population Fund 2014).[5] In this section, I will focus on the moral significance of the unequal benefits that accrue to people today and in the future due to the consequences of past emissions-generating activities.

Should the beneficial consequences of historical emissions be considered relevant to the determination of the fair distribution of the remaining permissible emissions among those currently living? If so, how? We can distinguish between several main objections against taking into account past emissions:

1. The United States, for example, caused over half of its emissions before 1975.[6] Posner and Weisbach (2007) point out that half of all American citizens living today were born after 1975 and over 27 percent are younger than twenty years old (ibid., 103). These young Americans could object: "Why should we be responsible for our ancestors' misconduct?" This objection holds that people living today should not be held responsible for the actions of their ancestors and should not be disadvantaged simply because those who previously lived in their country have emitted too much.
2. Posner and Weisbach (2007) suggest that a distinction should be made between those greenhouse emissions that were produced before the problem of man-made climate change was well known and widely recognized and those emissions caused later (ibid., 104, 111). For the period before widespread knowledge, many could object: "We didn't know about the greenhouse effect." This objection states that people can only be held morally responsible for actions that harm others if they knew or should have known about the potential for harmful consequences of their actions. It seems doubtful that most people before 1990, for example, could have known about the harmful effects of emissions (cf. ibid., 104, 110–16; see the discussion of the "culpability problem").
3. A third general objection appeals to the consequences of the nonidentity problem for the claim that past emission-generating activities have made some presently living or future people worse-off. This is because had a different environmental policy been pursued, the people living today most likely would never have existed as people with exactly their personal identity.[7]

The objections concern different subsets of past emissions. The first objection is relevant for emissions by people who are no longer alive; the second objection concerns emissions that were (presumably) caused before the publication of the first assessment report of the Intergovernmental Panel on Climate Change (IPCC) in 1990;[8] and the third can be raised regarding emissions (and regulations that affect emissions) that happened far enough in the past to be a necessary condition of the personal identity of people living today.

Despite these objections, I argue that at least some historical emissions should be seen as relevant for determining the fair distribution of the remaining emissions budget among people living today and in the future. This is because we are concerned with the fair distribution of the beneficial consequences of emissions-generating activities and not the distribution of emissions as such. In the following, I will argue that the three objections do not speak against taking into account, first, the unequal benefits people have realized from their own emissions-generating activities, at least since the time at which they can be held morally responsible for causing emissions. Second, they do not speak against taking into account the emissions-generating activities of those living earlier, insofar as they have unequal beneficial consequences for those living today and in the future.

The first proposal to take past emissions into account when allocating the remaining emissions among countries is based on the idea of fairly taking into account the benefits people have received from emissions-generating activities at least since they were liable to know that these activities contributed to the problem of man-made climate change. To get to this point, let us assume that the distribution principle applicable here is the EPC distribution of emissions rights.[9] One might think that people should have the same number of emission rights every day. Yet, this principle of distribution faces issues. First, the fact that people do not cause emissions from time to time speaks in favor of an even distribution over the entire lifespan of people (see Holtung and Rasmussen 2007; Hurka 1993, 9–22). The need to cause emissions does not only arise sporadically but also people cannot avoid emissions-generating activities. Today, emissions are necessary for almost all activities and in all phases of life. This will continue to be the case for the foreseeable future, and so far there has been a strong correlation between levels of emissions and benefits realized, if the accounting of emissions is done in a consumption-based way.[10] According to the broad consensus among climate scientists, the transformation to an almost emissions-free economy and society must be achieved by 2050 in order to avert the consequences of so-called "dangerous climate change" that are expected to come with an increase in the global average temperature of over 1.5° or 2° Celsius (see The Paris Agreement, Article 4 in conjunction with Article 2 [UNFCC 2015]). This transformation assumes either that greenhouse gases are avoided to a large extent or that they are compensated for or extracted after they have been generated. The costs that this transformation entail—not just for individual states and regions but also for individuals—are likely to depend heavily on how many emissions these actors will still be allowed to cause in the process of transformation. This means that the emissions budget (and each state's current emission level) determine how high the average reduction rate must be in order to achieve the goal of an emission-neutral future in about three decades, and this will in turn determine how drastic the reduction measures and how extensive the compensation and extraction measures must be in order to reach that goal.

If the allocation of future emissions rights relates to a fair share of the remaining emissions budget over the entire lifetime of individuals, then the emissions that currently living people caused and for which they can be held liable must be included. Apparently, this is only a part of all emissions caused by humans, namely ca. 11 percent (taking into account currently living people older than eighteen between 1990 and 2014).[11] However, the benefits from these emissions-generating activities have been unequally distributed among people, namely depending on how much benefits people have been able to realize from their own and other people's emission-generating activities. If a fair share of emissions stands for a fair share of benefits from emissions-generating activities, people in highly industrialized countries have typically already realized a large amount of benefits from their own liable emissions-generating activities. So if we want to achieve an EPC distribution of benefits, the following applies: a large proportion of the permits for beneficial emissions-generating activities would have to go to those who live in developing countries and have typically enjoyed far fewer such advantages in their lives. This is a way to argue for above-average per capita emissions rights for most people in developing countries. The argument is based on the inequality of the benefits from liable emissions-generating activities that people living today have produced over the course of their lives.

The first and third objections reviewed above do not speak against this position, but the second objection may. People may have benefited from their past emissions while being blamelessly ignorant of the consequences of these actions for climate change, for example, before 1990 (see footnote 7 above). Insofar as they could not know that their emissions-generating activities contributed to climate change, nor are they to blame for their ignorance, they cannot be blamed for benefiting from those emissions. Thus, at least for the purpose of my argument here,[12] these emissions and their beneficiary consequences should be considered irrelevant for the distribution of the remaining permissible global budget of emissions.[13]

Having shown the normative significance of the unequal benefits people have realized from their own past emissions-generating activities (at least insofar as they can be understood to have been liable for them), I now turn to the second proposal for taking into account the consequences of some historical emissions in allocating the remaining permissible GCB. The second proposal is that the emissions-generating activities of other people, including past people that have benefited currently living people, can and should be taken into account in the calculation of a fair distribution of the remaining

GCB and, thus, of an important constraint how states can undergo the transformation to net zero carbon emissions by 2050. To the extent that inherited benefits undermine what can be considered a fair distribution of benefits from emissions-generating activities they should be considered as undeserved benefits.

The industrialization pursued by our predecessors has had beneficial effects to this day, and this benefit has been greater for people in highly industrialized countries than for those who live in developing or emerging countries. These benefits include, for example, the provision of complex infrastructures including educational facilities, hospitals, roads, and train connections, which were often set up and built before those currently alive were born. In this way, further benefits of past or historical emissions (ca. 5 percent of all historical emissions [from 1850 to 1990])[14] are taken into account for the fair distribution of the remaining emissions budget, even beyond those covered in the first proposal.

The basic elements of the argument for taking into account these benefits are the following: first, we should think about the GCB not in terms of emissions, but in terms of benefits from emissions-generating activities.[15] Second, the unequal inherited benefits from previously living people's emissions-generating activities make a difference in terms of the benefits currently living people have realized and will realize during their lifetime. Third, these pre-1990 emissions-generating activities can be considered wrongless with respect to the emissions caused, yet are still relevant. Fourth, currently living people cannot have done anything to deserve a higher share of these benefits. Fifth, we can rely on a principle of distribution to determine the fair distribution of benefits from emissions-generating activities over the lifetime of people. Sixth, as I have argued elsewhere, the EPC principle for distributing "benefits from emission-generating activities over the whole life-time of individuals" can be justified on egalitarian and nonegalitarian grounds (Meyer and Roser 2006, 238–40). Seventh, as the unequal consequences of previously living people's wrongless emissions-generating activities are due to luck and higher shares of these benefits are undeserved, distributive justice requires taking them into account in specifying the allocation of the remaining emissions budget in terms of permissions to perform emissions-generating activities that have the potential to benefit those who engage in these activities.

As such, taking into account inherited benefits is not open to the three objections. Doing so does not imply holding those living today responsible for the emissions-generating actions of those living earlier. Rather, the notion of inherited benefits considers these consequences as a matter of luck. Moreover, taking into account these benefits is fully compatible with those who produced them being excusably ignorant about the consequences of causing emissions.

Consequently, the first two of the above-named objections do not speak against taking this part of historical emissions into account. The third objection, which relies on the nonidentity problem, seems to speak against taking into account the received benefits from past people's actions. On closer inspection, however, this is not the case. This is because this argument is based on a view of distributive justice and does not depend on any claim that present people are better off or worse off than they would have been had history taken a different course. The nonidentity problem arises because no one can say that he or she has been benefited or harmed by industrialization, since, given a different history of economic development, they most likely would not have come into existence. This problem does not undermine the claim that the circumstances in which people have been raised and lived in since conception are more or less beneficial to them. People receive benefits depending on whether they grew up in the industrialized world. An individual would obviously have fared differently if she had been moved to a slum in one of the developing countries soon after her birth in a highly industrialized country. This approach takes into account that these benefits are in part the result of the emissions-generating activities of their predecessors.

To summarize, certain parts of past emissions should therefore be taken into account for the fair distribution of the remaining emissions budget. First, the past emissions that are the result of activities that have benefited present people since 1990 matter, according to the arguments above. Second, those past emissions that are the result of emissions-generating activities of others that still benefit people living today and in the future likewise matter.[16] Accordingly, the inherited benefits argument suggests the following benefits are relevant: first, the unequal benefits that currently living people have realized at least since they can be considered adults[17] through their own emissions-generating activities; and second, the unequal net benefits that living people have realized since their conception due to the (beneficial and harmful) long-term consequences of the emissions-generating activities of previously living people.[18]

These results are relevant for determining states' fair share of the GCB. First, states ought to be allocated a smaller share of the remaining emissions budget if, since 1990, their populations have realized greater than average net benefits from their own emissions-generating activities. Second, states ought to be allocated a smaller share of the remaining emissions budget if their populations have benefited from the emissions-generating activities of past people more than on average. Of course, states whose populations have realized less benefits from past and historical emissions have a justified claim to a larger share of the remaining permissible global budget of emissions. (In section "Feasibility Concerns," I will introduce ways of operationalizing these considerations of historical responsibility and discuss the feasibility concerns regarding the distributive implications of this view.)

Since these two arguments only take part of past emissions into account, large inequalities in the generation of emissions among past people are not taken into account. For example, historical emissions that can be traced back to the emissions-generating activities of past people and that benefited them but not people currently living are without normative significance. They amount to ca. 61 percent of total historical emissions.[19]

To conclude, if we understand the distribution of emissions rights as a problem of distributive justice, we are not referring to harm or wrongdoing. Rather, the goal is to distribute the benefits associated with emissions-generating activities fairly among those currently living over their whole lifetime. As argued, we will not achieve this goal if we do not give people from developing countries greater shares of emissions rights. This is due to the unequal inheritance of benefits, as well as the unequal benefits that people could achieve through their own emissions-generating actions.[20]

While important, I will not now determine the exact implications of this view of the significance of historical emissions (see Williges et al. under review). Rather, my aim is to show the normative relevance of historical emissions in terms of the past and present benefits that they have provided. These normative considerations, however, may actually justify different claims for the distribution of the remaining permissible emissions budget. If so, consistency then seems to require that we consider all normatively relevant claims when allocating the emissions budget.

II. THE NORMATIVE SIGNIFICANCE OF HISTORICALLY SHAPED LEGITIMATE EXPECTATIONS

In this section, I will examine one such claim. This is the view that high emitters are owed a larger share of emissions rights due to their historically developed "legitimate expectations." According to this argument, the expectations that historically high emitters have developed—that is, to be allowed to emit at levels well above average—are not to be attributed to them, but to the historical development of emissions levels within the highly industrialized countries, for which they are not responsible. And owing to these expectations, if historically high emitters are asked to quickly and drastically reduce their emissions (specifically to net zero), they will have to incur higher costs than people who strive to achieve this goal starting from a lower emissions level. These higher costs are relevant, so the argument goes, for the fair distribution of the benefits from emissions-generating activities.

Expectations can be understood as a certain type of prediction about the future that is characterized by three features. First, expectations play a major role in people's life planning and in the execution of their projects. Expectations are part of the background against which agents choose between different (long-term) projects. In the context of the discussion of this chapter, I am concerned with the expectations surrounding the level at which agents will be able to afford to cause emissions or at which they may permissibly cause emissions in the pursuit of their projects. The pursuit of each of those long-term projects is associated with a certain level of emissions as an (so far) unavoidable side effect. If any of the potential long-term projects are associated with a level of emissions that is above the permissible level, then the agent has reasons to doubt that completing those projects will be possible. Likewise, after having engaged in a project with the expectation that its unavoidable level of emissions is permissible and affordable, people are likely to be harmed if this expectation is not met because less emissions-intensive substitutes may not be available and, thus, it will be difficult or impossible for them to continue to pursue their projects.

Second, the fulfillment or nonfulfillment of expectations is in principle under human control; whether expectations are met depends on people's actions and omissions. In the case under consideration, it is mainly a matter of future government policy whether people's expectations about their future affordable or permissible levels of emissions are fulfilled or frustrated. Third, the following discussion assumes that the expectations examined are epistemically valid, in the minimal sense that people who have them also have good reasons to believe that they will be fulfilled. For instance, the expectation that the government will raise the salary of teachers in public schools three times as much as that of judges and police officers is evidently epistemically invalid under normal conditions. If this were an expectation of a teacher, we would kindly characterize it as wishful thinking.

Because of these features, expectations and the actions of people that affect their fulfillment or nonfulfillment may be subject to normative assessment. For example, do people have valid claims that their expectations will be met? Should other people refrain from undermining these expectations? What counts as a *legitimate* expectation? What claims do people have when their legitimate expectations are frustrated?

The following two examples can illustrate the difference between legitimate and illegitimate expectations. The first represents an intuitively illegitimate expectation. Consider a thief who steals a car and expects to get away with the theft. Because he has this expectation, he makes various financial commitments getting out of which will cost him or harm him. Let's say the police catch him and his expectation gets frustrated before he can benefit from the theft. He was certainly harmed by the frustration of his expectation, but intuitively we would not believe that he has any kind of claim to compensation or any kind of valid complaint in general. In this case, the fact that an actor has a certain expectation and would be harmed by its frustration does not create a valid normative claim by the actor not to be harmed that way.

Let us now consider the second case. This case concerns a cooperation based on trust and reliability. For some time now, two roommates A and B have dinner together on Fridays and alternately prepare dinner. Suppose A prepares dinner this Friday because it was her turn. A relies on her expectation that B will be home for dinner. But B does not show up. A is frustrated. Indeed, A suffers harm due to the frustration of her expectation. Although the frustration of A's expectation is relatively insignificant, one would normally assume that B owes her something, at least an explanation, and probably also an apology. In this way, A's expectation is normatively significant and thus legitimate. If it becomes frustrated, B should react, and A has reason to complain that her expectation has been frustrated.

We can assume that the first case represents an illegitimate expectation and the second case a legitimate expectation (but for a more complex understanding, see Meyer and Sanklecha 2014). Furthermore, it is assumed that one of the implications of a legitimate expectation is that its bearer has a valid normative claim that the harm that can result from the frustration of the expectation is taken into account when decisions and actions are taken that can lead to its frustration.

It is not controversial that people living in countries with a high level of per capita emissions typically have a wide range of expectations that are likely to be frustrated if climate change is dealt with appropriately. Think of expectations regarding the level of future permissible personal emissions in private activities (e.g., emissions from the private use of cars with combustion engines), expectations regarding future demand for the skills in which people were trained (e.g., when they were trained to become automechanics and can maintain and repair internal combustion engines), expectations about the future permissibility and affordability of long-haul flights (e.g., if people are studying on another continent but regularly want to fly home for personal reasons), among others. In addition, and generally speaking, these people will have expectations that they will be able to maintain and pursue their current ways of life in the future. This is part of the background against which actors choose and pursue long-term plans and projects, and the ability to do so is an integral part of the good life (Rawls 1999, 358–60; Williams 1973, 116–17).

Previous studies have examined whether, and under what conditions, expectations that are based on continuing activities that cause a relatively high level of emissions are legitimate, and what this would mean normatively and practically speaking (Meyer and Sanklecha 2011; Meyer and Sanklecha 2014; Ortner et al. 2017). However, the situation is complex. On the one hand, global emissions need to be reduced significantly if the temperature rise is to be kept below 2° or 1.5°, and it will generally not be possible for people who emit at a particularly high level today to continue to emit at their current level if these reductions are implemented. In addition, fairness considerations seem to require that current high emitters are responsible for at least some of the burden of reduction. Put succinctly, we can say that intergenerational justice considerations support the demand to significantly reduce global emissions, and considerations of international distributive justice support the demand that the burdens of this transformation to net zero emissions are predominantly born by those engaged in high-emissions activities.

On the other hand, we are faced with the following considerations. First, the actors whose expectations are frustrated will often suffer harm. In some cases, the damage can be significant as it can lead to the actor having to give up long-term projects and plans in which he or she has already invested heavily (and not only financially). Unfortunately, to achieve an intergenerationally just reduction in global emissions, the average reduction rate must be large and over a short period of time. Thus, there will be high costs connected to the frustration of the status quo-expectations of high emitters and the associated changes in their ways of life that are necessary to reduce their emissions to net zero. For the sake of the argument, let us also assume that these personal costs will be higher than for people starting from a much lower level of personal emissions activities.

Second, in many cases, it will be difficult to hold actors responsible for the expectations, plans, or projects that they have developed or pursued. This is because people developed their long-term plans and projects based on the options that are typically and realistically available in high-emitting countries. If such expectations meet certain conditions, at least some of them can be considered legitimate.[21] Even if the historical processes that were constitutive for the formation of these expectations are considered wrongful today,[22] the people whose expectations were formed by these processes are not responsible for them. This is because they could not influence activities before their birth or have had a negligible influence on these processes and their development and institutionalization.

However, the fact that an expectation is legitimate does not mean that it must be protected for any reason. In the context of climate change, there are important reasons for frustrating high emitters' legitimate expectations to continue to cause emissions far beyond their EPC allocation. As outlined above, protecting the expectations of high emitters would be inconsistent with fundamental considerations of intergenerational and international justice. Global emissions must be reduced significantly, and this reduced GCB should be fairly distributed among the states. In other words, that the expectations of high emitters are legitimate should generally not result in any normative claims that have more weight than the fulfillment of the duties of intergenerational and international justice. Of course, judging a claim as less weighty or as subordinate to other normative requirements does not mean that the claim is to be ignored (or that it is rightly ignored). Rather, certain emissions expectations are legitimate and should not be ignored.

The understanding put forward here can therefore be summarized as follows: the frustration of legitimate expectations can be regarded as permissible, since the reasons for protecting legitimate expectations have to be weighed against considerations of international and intergenerational justice. To be sure, the permissible global budget is limited by the rights of future people vis-à-vis currently living people (in particular the sufficiency rights of those living in the future) (Meyer 2009; Meyer and Pölzler forthcoming). The international allocation of the remaining budget ought to reflect the EPC principle. At the same time, the frustration of a legitimate expectation in the context of climate change leads to a *pro tanto* valid claim that these burdens be taken into account when the remaining permissible emissions are distributed among people. According to the argument of legitimate expectations, the following costs have to be taken into account from the point of view of fairness: the actual and expected adaptation and substitution costs of currently living people due to the frustration of historically shaped legitimate expectations, and the losses in the event that the projects cannot be continued. When the personal costs associated with reducing emissions to net zero born by high emitters can be shown to be (possibly far) higher than average, then this matters for the fair allocation of benefits from emissions-generating activities over the whole lifetimes of people, and so these ought to be taken into account. Having to bear above-average costs can justify a higher share of the remaining permissible emissions. In this way, the argument of legitimate expectations can be understood to justify some (highly limited) form of grandfathering.[23]

III. HOW THE ARGUMENTS ARE SIMILAR

Both arguments—the argument regarding inherited net benefits and the argument regarding legitimate expectations—are based on the idea that we should be interested in both the beneficial and harmful consequences of historical emissions-generating activities (and not in the emissions themselves). In the following sections of this chapter, I will argue that coherence requires that whoever supports the inherited benefit argument should also support the legitimate expectation argument (and *vice versa*).[24] At the same time, the two arguments have different normative implications.

Both arguments are relevantly similar. As stated above, when it comes to inherited benefits, there are serious difficulties in identifying misconduct with regard to historical and past emissions. However, distributive justice requires us to react to the highly unequal consequences of historical and past emissions, even if no wrongful harm has been committed. The argument in favor of taking inherited benefits into account presupposes that people can be considered disadvantaged or privileged due to the emissions-generating activities of past people without deserving this inequality (and without being responsible for it). The resulting distribution is thus unfair or unjust according to an outcome-oriented principle of equity. Accordingly, people stand under obligations to redistribute the relevant goods with the aim of achieving a just distribution.

When it comes to legitimate expectations, their frustration will often be harmful. Forgoing long-term plans and projects that are connected to legitimate expectations can typically cause significant harm (in terms of the costs of adaptation, substitution, and loss of valuable goods). But it is nonetheless permissible—if not obligatory—for governments to frustrate them in the course of implementing a just regime of transformation to net zero emissions. However, the harm remains normatively relevant and harms owing to the frustration of legitimate status quo-expectations are among the net costs that people incur in the transition to a low-emission society. These harms must be taken into account.

We can see that both arguments depend on the assumption that the consequences of historical emissions-generating activities should be dealt with from the point of view of distributive justice. They are also both arguments based on the idea that currently living people should not be held responsible for at least some of the consequences of historical emissions. Specifically, the inherited benefit argument argues that people living today are neither responsible for nor deserve the net benefits they enjoy owing to unequal historical emissions. Likewise, the argument regarding legitimate expectations concludes that people living today are not responsible for at least some of the expectations they have developed due to historically high emissions, and therefore do not deserve the harm they will suffer from the frustration of these expectations.

In summary, the inherited benefits argument and the argument regarding the harms due to the frustration of legitimate expectations concur that the unequal beneficial and harmful consequences of past emissions-generating activities are relevant for reasons of distributive justice, but each of the arguments focuses on something different. While the inherited benefit argument is about the unequal positive consequences of past emissions for the well-being of those living today, the argument of legitimate expectations focuses entirely on the unequal harmful consequences that are expected to arise in transforming from a high-emission to a low-emission regime through the frustration of legitimate status quo-expectations. Because the arguments are based on the same basic rationale—that distributive justice demands evening out unequal undeserved benefits and burdens—coherence requires recognizing the validity of both arguments.

IV. WHY THE ARGUMENTS HAVE DIFFERENT NORMATIVE IMPLICATIONS

One might think that the legitimate emissions expectations of high emitters could justify an increase in their share of the global emissions budget. For example, a temporary grandfathering of their levels of emissions could take into account high emitters' normatively relevant costs. These levels would be consistent with the equal emissions per capita approach to the transformation to net zero emissions by 2050. Instead of immediately implementing EPC shares of the remaining global emissions budget, the legitimate expectations of individuals in countries with particularly high emissions would be seen as a reason for allocating the remaining global emissions budget in accordance with a "contraction and convergence" approach. This approach reduces emissions levels based on different actual starting levels with a gradual transition to the same net zero-per-capita distribution of emissions by 2050 (the implications of this approach are discussed in Williges et al. under review).

As I reported in the introduction, states' commitments under the Paris Agreement are heavily dependent on their historically achieved level of emissions; high-emitting states implicitly claim for themselves grandfathering rights. However, I will argue that the argument from legitimate status quo expectations cannot support grandfathering in terms of an increased share of the remaining permissible global budget. Rather, the above-average costs of the frustration of legitimate status quo expectations are distributively relevant in a different way, namely in terms of the fair distribution of the costs of transformation to net zero.

Let me first note that it is a rather difficult empirical issue to determine who has above-average costs when legitimate status quo expectations are being frustrated. It seems questionable that reducing emissions from a high to a very low level in a short time necessarily comes with higher costs than those that low-emitting nations will bear to reach the same level in the same amount of time. Indeed, it may actually be less costly for countries with far above-average emission levels to reduce emissions than for low-emitting countries to reduce to the same level in a short period of time. For instance, because of the positive correlation between (historical) emissions and welfare levels, states with emissions levels far above average are more likely to be able to afford to implement adaptation measures and accordingly have more and better options in terms of avoiding substitution costs and losses. Also, high-emitting countries will tend to have fewer losses and damages as a result of climate change, as well as lower adaptation costs due to a lower percentage of especially welfare-relevant sectors that are particularly affected by the consequences of climate change (e.g., agriculture). Due to such factors, countries with above-average emissions levels are likely to have lower collective costs and therefore their residents will have lower reduction costs on average.

Furthermore, even if residents of countries with well above-average emission levels face special and higher costs due to the frustration of their legitimate status quo expectations, this should not affect the allocation of the global budget for emissions rights. After all, not all actors and not all states are equally responsible for shaping the problematic status quo expectations. The highly industrialized states have played a constitutive role in shaping these expectations and continue to do so. States have a significant impact on what lifestyles are typically found in a given country. Since the state promotes some lifestyles and long-term projects (at the expense of others), one can say that it (a) is involved in a broad sense in determining the range of options from which its citizens choose their lifestyles and longer-term projects, and (b) it encourages its members to develop and hold the expectation that they will continue to be supported in pursuing their longer-term projects (including the level of emissions associated with them) (cf. Rawls 2005, 260, 270; Meyer 2015, 151–55).

In addition, since 1990 at the latest, states have had a responsibility to pursue a transformation to a low-carbon society, which comes with a change in the expectations of the members of these states with regard to a permissible level of emissions. Yet, most high-emitting states have failed to do so. States have the authority and ability to pursue such a strategy. States can take steps to determine how their residents can reduce their emissions while still being able to pursue their projects. The state can prohibit certain types of action and create incentives for changing behavior. It can influence the costs of reducing emissions associated with the pursuit of individual conceptions of the good, either by reducing the costs of less emissions-intensive means of, for example, energy production or mobility (which are relevant to the pursuit of most projects) or by subsidizing certain types of projects, such as a specific form of agriculture. It can invest in infrastructure that allows activities to be carried out with lower emissions costs than they would otherwise necessitate by, for example, creating convenient and efficient public transport. Finally, the state could also set an upper limit on personal emissions for private activities, thereby further influencing its members' expectations regarding emissions. Most high-emitting states are not currently doing so and are thus conveying the message to their citizens that any amount of personal emissions is allowed. To be sure, this affects the projects that its citizens select and their expectations of future emissions.

To summarize, there are three reasons to impose the costs of reducing emissions from a particularly high level to a net zero level on states with a well above-average level of emissions. First, the states with levels well above average are largely responsible for the development of this level, historically and causally. Second, the states themselves are able to take authoritative and effective measures to lower emissions levels in such a way that the reduction is as compatible as possible with the continuation of the lifestyles and projects of their residents. And third, for reasons of intergenerational and international justice, states with well above-average emissions levels have had, for some time, a duty to reduce emissions levels.

At the same time, it is likely that for some people the costs of reducing emissions to a significantly lower level is particularly high for reasons for which they are not responsible. For example, they may be pursuing a project that involves very high emissions, and for which a much-less emissions-intensive substitute is very expensive or unavailable. Also, measures implemented by the state to reduce emissions levels can impact people's options for being able to continue to pursue their projects very differently. This suggests that states should take into account such unequal consequences when choosing and pursuing their reduction strategies and that they should seek to avoid particularly high costs for individuals through the frustration of their legitimate status quo expectations, while providing measures of compensation to those unequally or seriously burdened. At the same time, however, the special historical and causal responsibility of states with far above-average emissions levels and their long-standing duty to meet their intergenerational and international obligations to reduce emissions speaks against imposing their supposedly particularly high costs of emissions reduction on everyone by allocating these high-emitting states a larger share of the remaining emissions rights.

While we have strong reasons to object to the notion that the legitimate expectations argument can justify historically and currently high-emitting states being allocated larger shares of the remaining permissible global budget, the inherited benefits argument typically does justify allocating (much) smaller shares to these states, as argued above. Specifically, low-emitting people typically have inherited (far) fewer benefits from previously living people's emissions-generating activities and have likewise realized (far) fewer benefits through their own emissions-generating actions. Consequently, low-emitting states ought to be allocated a (far) greater shares of emissions rights, in line with the goal of fairly distributing the benefits associated with emissions-generating activities among those currently living over their whole lifetime.

My analysis of the significance of historical emissions shows that only the inherited benefit argument (and not the legitimate expectations argument) is relevant to the fair allocation of the remaining permissible global budget of emissions. It demands that we decrease the shares allocated to (historically and currently) high-emitting states and increase the shares allocated to low-emitting states. In (implicitly) assuming grandfathering rights, representatives of high-emitting states might mistakenly believe that their citizens' supposed special costs of reducing to net zero emissions justify higher shares of the global budget for high-emitting states and, thus, would counterbalance the reduction of their national budgets that is required owing to them having inherited far more benefits from past emissions-generating activities. However, properly understood, as argued here and in the previous two sections, the special costs of some high emitters caused by the frustration of legitimate expectations can, at most, justify a plausible criterion of fairness for intra-national net zero transition policies.

V. FEASIBILITY CONCERNS

My analysis of the significance of historical emissions so far has not weakened the requirements for highly industrialized countries with regard to their emissions reductions. To the contrary, it has fortified these requirements by showing that high-emitting states' shares of the global emissions budget are to be reduced owing to their above-average inherited and realized benefits from historical and past emissions. To be sure, the argument as presented considers only some parts of historical emissions significant. Specifically, it recommends disregarding all pre-1990 emissions that have no unequal consequences for the well-being of currently living and future people, as they are irrelevant to the fair allocation of benefits from emissions-generating activities among current and future people. However, if a just climate policy is to reflect the allocation of emissions rights in accordance with the arguments above (i.e., fewer emissions rights for traditionally high-emitting countries, and a greater amount of emissions rights for traditionally low-emitting countries), then we may wonder whether and to what extent the adoption of such a policy is feasible. I will argue for a diachronically dynamic view of feasibility according to which agents, if so committed, can overcome (so-called "soft") feasibility constraints in time and this includes high emitters overcoming constraints that restrict their ability to reduce emissions.

Intergenerational justice requires limiting global emissions in a way that is compatible with staying well below a 2° rise in global average temperature. Thus, reaching an effective international agreement on how to distribute the resulting GCB is urgent. A failure to do so will make it very likely that we will have a far higher temperature increase, and thus risk dangerous climate change that would seriously harm the fundamental interests and violate the basic rights of very many people living in the future. To prevent this, the agreement would have to aim at bringing global emissions down to net zero by 2050. At the same time, an effective international agreement will also have to define an allocation of the remaining global emissions budget among states. According to Williges et al. (under review), taking into account historical emissions in the two ways I have argued are required for reasons of distributive justice would both result in a decrease of the national budgets of highly industrialized states. But both ways of taking historical emissions into account cannot be criticized in terms of normative justification. Rather, they can only be criticized as having infeasible implications.

Highly industrialized states tend to favor a contraction and convergence (CAC) approach (ibid.). According to CAC, every country begins with its current average per person emissions and converges on a globally common future level of per-person emissions by a future point in time, namely 2050. CAC implies a strong form of grandfathering: the higher the current levels of emissions, the higher the share of the GCB. CAC takes these unequal levels as the baseline for the allocation of the GCB to the individual countries. CAC does not take into account past and historical emissions. As a result, highly industrialized countries will have more emissions rights during the transition phase to a low-carbon society, and so will have more options for implementing the transformation and less associated costs.

According to the alternative EPC approach, the GCB should be split in such a way that all countries are allocated an equal amount of emissions per person from today until 2050. EPC does not imply grandfathering, nor does it take the current levels of emissions for granted or take currently reached levels as relevant for the allocation of the GCB. EPC reflects the idea of an equal distribution of the remaining permissible emissions among persons and from now on. However, both the simple understandings of EPC and CAC disregard the past. Neither accounts for past emissions or the consequences of pre-1990 emissions.

In the following, I will distinguish between six understandings of the allocation of the GCB: simple (historically blind) CAC and EPC and qualified understandings of CAC and EPC. I will refer to those views that take into account post-1990 emissions as limited historically sensitive CAC and EPC; similarly, I will refer to those views that consider both post-1990 emissions and inherited benefits as fully historically sensitive CAC and EPC. As argued in the previous sections of this chapter (a specific interpretation of) the fully historically sensitive EPC is justified and is morally superior to the other five understandings of the allocation.[25]

However (and interestingly), Williges et al. (under review) show that a fully historically sensitive CAC approach turns out to be similar to an allocation based on a simple, historically blind EPC.

To review, the first way that fair historical responsibility requires that we take historical and past emissions into account was by attributing responsibility for emissions to emitters from 1990 onwards. This would require us to take 1990 as the starting point for the process of convergence to the net zero contraction point. The second way that fair historical responsibility requires we take historical and past emissions into account was by taking into account inherited benefits. As a proxy for inherited benefits, Williges et al. (under review) propose to estimate the carbon emissions embodied in infrastructure available in 1990. These capital stock emissions are added to the initial GCB. This larger budget is then distributed to countries, minus their individual embodied emissions. In that way we account for the unequal consequences of presumably wrongless past emissions (that agents caused while being nonculpably ignorant about their adverse consequences) on the basis of the principle of equal benefits from emissions-generating activities for currently living and future people. Historical and past emissions understood in these two ways turn out to be so high that they account for much of the difference between the international allocations recommended by the simple CAC and EPC approaches. Thus, if qualified in these two ways the CAC approach leads to an allocation of the remaining global budget that is very similar to the allocation determined by the (historically blind) EPC approach.

On the basis of my analysis of the normative significance of historical emissions we should reject the simple CAC approach as clearly unjustified and unfair owing to the fact that there is no justification for taking current levels of emissions for granted and making them relevant to the distribution of the remaining global budget. Qualifying the approach by taking into account historical emissions significantly decreases the questionable grandfathering character of the CAC approach. The full historically sensitive CAC approach clearly is morally superior to the simple CAC approach. Of course, the full historically sensitive CAC approach could be criticized for having infeasible implications. However, if the resulting allocation turns out to be feasible—in particular, for highly industrialized states—then there are strong reasons to favor the implementation of the full historically sensitive CAC.

The CAC approach can be understood to be historically sensitive also in the limited way. Other authors have argued for taking into account only past emissions since 1990 (Kenehan 2017, 207–12). Emissions from 1990 onwards represent 38 percent of all emissions from 1850 to today.[26] Two reasons seem to speak in favor of adopting the limited historically sensitive CAC: first, taking into account inherited benefits is subject to specific normative objections as the idea of counting inherited benefits raises justificatory issues having to do with the relevance of emissions-generating activities of mostly previously living people. In the section "The Normative Significance of Having Received Benefits from Emissions-Generating Activities," I argued that taking into account inherited benefits from a distributive justice perspective can be shown to be immune to such objections. However, I would like to acknowledge a second concern: taking into account inherited benefits does raise comparatively difficult issues of operationalization. The proposed proxy of capital stock for inherited benefits from past emissions-generating activities is open to objections, for example, the objection that the inherited infrastructure might actually be a burden in terms of achieving emissions reductions. In any case, while responsibility for emissions since 1990 is widely agreed to be both plausible and straightforwardly operationalizable, there is less agreement on the justifiability of historical responsibility for inherited benefits, and its operationalization is more difficult.

As taking into account both post-1990 emissions and inherited benefits from pre-1990 emissions-generating activities will reduce the carbon budgets of highly industrialized states, taking into account only the former, that is the post-1990 emissions, will allocate highly industrialized states a higher share of the remaining GCB. Thus, the limited historically sensitive CAC likely leads to a more feasible allocation of the GCB for highly industrialized states (ibid., 204–6) than the full historically limited CAC. However, the limited historically sensitive CAC keeps some of its strong grandfathering character intact and, thus, the approach is open to the charge of unjustified discrimination in favor of the undeserved beneficiaries of the problem of climate change. Qualifying CAC by taking into account inherited benefits is well justified, and doing so will further decrease the strength of the unjustified grandfathering of the CAC approach, and will increase the acceptability of the CAC approach from the perspective of developing countries.

Generally speaking, developing countries favor the EPC approach. Thus, the full historically sensitive CAC approach (that, as noted before, contingently amounts to much the same in terms of the allocation of the global budget among states as the simple (historically blind) EPC approach) is clearly better justified and more acceptable as well as more feasible for developing countries. For reasons of justifiability, we should aim at realizing an agreement that reflects both ways of taking historical emissions into account.

A full historically sensitive CAC approach might seem less feasible today than the limited historically sensitive CAC. Such a comparative claim relies on an assessment of the negotiation power of the high-emitting countries and supposes that they will use their power to push for the allocation that is less burdensome for them and also, incidentally, less just. However, as Brennan and Sayre-McCord argue (2016, 442), "the fact that one option [. . .] is normatively better than another constitutes a consideration that might make achieving it more likely to be feasible (given the sacrifices involved) than it would otherwise be. If people care about justice and can be convinced that [one option] is more just, those people, at least, will be more willing to make sacrifices and cast votes and change their behavior in other ways that might make [that option] feasible." In any case, based on the arguments in the section "The Normative Significance of Having Received Benefits from Emissions-Generating Activities," we have strong reasons to work toward changing the feasibility constraints that hinder our ability to take into account inherited benefits.

Not considering questions of feasibility, in this chapter and in Williges et al. (under review), I argue that a fully historically sensitive EPC approach is normatively best justified. The resulting allocation is ideally fair and ideally justified. However, such an allocation is most likely not feasible for highly industrialized states under current conditions. Arguably such an allocation of the remaining permissible emissions is infeasible in the short run, namely under current conditions. However, it is possibly dynamically feasible to an increasing degree if a transformation of the relevant (so-called "soft") constraints can be achieved (Gilabert and Lawford-Smith 2012, 814–16; Lawford-Smith 2013, 252–56; Jewell and Cherp 2020, 6–8).

The feasibility of such an allocation, and thus its accompanying average reduction rate for states, might be increased. The feasible reduction rate depends on cultural, technological, and institutional constraints that can be changed. A larger reduction is possible when people either realize welfare in a less emissions-intensive way than had so far been assumed to be likely or when people change their ways of life so that they need less emissions for realizing well-being (newly understood). As suggested above, people's normative conviction that they ought to support such changes can make these changes more feasible. When it comes to the technological constraints, a faster-than-expected increase of the share of renewable energy resources will decrease emission intensity and the emissions budget of countries might be increased if the technology of carbon storage becomes available to a larger extent than so far assumed. Similarly, if the efficient distribution of emissions rights by, for example, a global market of tradable emissions rights can be institutionalized (a process currently facing institutional constraints), highly industrialized countries are likely to be able to buy emissions permits so that they may initially reduce their emissions less radically.

In summary—and taking seriously both justifiability and feasibility—policymakers should aim at reaching an international agreement that allocates the GCB on the basis of either the EPC approach or a historically sensitive CAC approach, taking post-1990 emissions into account or preferably taking into account both post-1990 emissions and inherited benefits. The latter, the full historically sensitive CAC approach, amounts to an allocation of the global budget among states that is similar to the allocation specified by the historically blind EPC approach. For different reasons, states can consider either of these two approaches, the EPC approach and the full historically sensitive CAC approach, as desirable and justified under current conditions of feasibility.

This amounts to proposing that, for reasons of feasibility, we change the weight given to considerations of historical responsibility for the time being (namely by considering the two ways of taking historical emissions into account as qualifications of the CAC approach and not of the EPC approach). By doing so, one accepts (for reasons of feasibility) that highly industrialized states have a higher share of the global budget than justice ideally requires. However, given the diachronically dynamic character of the relevance of feasibility considerations (Gilabert 2012, 47–52), this is not the end of the story.

By adopting a dynamic understanding of feasibility, the international agreement should oblige all states to work toward changing the relevant constraints in such ways that a historically sensitive EPC approach might become feasible. High-emitting states should oblige themselves to pursue policies that aim at changing the cultural, technological, and institutional constraints that currently limit their capacities to reduce emissions. Their efforts should be monitored, and success in overcoming the constraints should be regularly reviewed as part of the agreement. If the transformation is successful, this could allow for the incremental consideration of historical emissions under EPC, until we realize an allocation of the global budget that reflects what the full historically sensitive EPC approach specifies.

Unless high-emitting states seriously commit to transforming the constraints so that they will become able to realize more demanding options of taking into account historical emissions (that is making these morally required options feasible for them), the high-emitting states can be understood as pursuing a strategy of extortion (Gardiner 2016, 90–9): assuming that states will want to agree on a global strategy of staying well below a certain threshold of temperature increase (e.g., "well below 2° Celsius"), the longer high-emitting states hold off on the implementation of demanding reduction rates that reflect their historical responsibility the less feasible the urgently needed limitation of the global budget of emissions is unless other states bear unfair reduction burdens.[27]

Here we face a problem of perverse incentives (Meyer and Waligore 2018, 230): high-emitting states have an incentive to remain incapable of fulfilling their duties concerning historical emissions as it seems likely that this will lead to a situation where states will finally accept an allocation of the costs of avoiding dangerous climate change that does not require high emitters to acknowledge any historical responsibility or even any historical responsibility to a plausible degree (a strategy apparently recommended by Weisbach

(2016, 226)). Those who will be wrongfully harmed by this outcome (by having their emissions budget unnecessarily and unfairly reduced) have strong reason to oppose such a development. In negotiating an international agreement on the allocation of the remaining permissible emissions, they should rightfully insist on a perspective of moving toward a more just distribution based on a credible commitment of high-emitting states to change the relevant feasibility constraints they currently face in reducing emissions drastically.

VI. CONCLUDING REMARKS

The consequences of past and historical emissions-generating activities can have strong normative significance from the point of view of distributive justice, if it can be shown that there are persons that are unequally affected in a distributively relevant way and that the persons concerned do not deserve this inequality. In such cases, these consequences can justify measures of evening out these differences. The normative implications of these unequal consequences can differ, depending on which consequences we analyze. In this chapter, I examined two subsets of these consequences, of which state representatives claim (implicitly and explicitly—see the introductory paragraphs of this chapter) that they are relevant to the allocation of the remaining permissible global emissions rights. My analysis has shown that this is true for one subset, but not for the other.

The unequal advantages owing to one's own past emissions-generating actions since 1990 and the unequal inherited benefits currently living and future people enjoy from pre-1990 emissions-generating activities of previously living people are relevant for the fair distribution of the remaining global emissions rights (see "The Normative Significance of Having Received Benefits from Emissions-Generating Activities"). This is so, because from the perspective of distributive justice (e.g., according to the principle of the EPC distribution of benefits from emissions-generating activities), the benefits of activities with emissions as a side-effect should be evened out over the whole lifespan of people. Given the significantly different amounts of benefits that have already been realized—through our own and past people's emissions-generating activities—a larger proportion of the remaining global emissions budget is to be allocated to those who have or will have realized fewer benefits. This is a matter of cosmopolitan or international justice.

Historical emissions-generating activities also have other consequences among them the status quo expectations of high emitters (see "The Normative Significance of Historically Shaped Legitimate Expectations"). For reasons of intergenerational justice, it will be necessary to frustrate many or most of these expectations. I argued that taking the costs of frustrating these expectations into account and providing measures of compensation for them is a matter of domestic distributive justice, not international justice. It is an empirical question and it is at least questionable whether highly industrialized countries with emissions levels far above average have special and particularly high costs in the transformation to net zero emissions by 2050, as explained in the section "Why the Arguments Have Different Normative Implications." In any case, over a long period of time the authoritative decisions and policies of the states with currently very high-emissions levels have been key in the formation of the status quo-expectations of their residents. With high historical and current levels of emissions, highly industrialized states have created the institutional framework for the emissions-intensive ways of life and options typical of these societies. Individual citizens choose from a range of lifestyles and options that they themselves have contributed little to generating. This goes hand-in-hand with status quo-expectations and, if qualified in a certain way, holding at least some of these expectations is not blameworthy (see "The Normative Significance of Historically Shaped Legitimate Expectations" and "How the Arguments Are Similar"). Also, due to the frustration of their legitimate status quo-expectations, people can have high and diverse costs (of adaptation, substitution, and loss) associated with a quick reduction to net zero.

People living today neither deserve the net benefits they enjoy owing to unequal historical emissions nor the harm they will suffer from the frustration of the historically formed expectations that they cannot be blamed for (see "How the Arguments Are Similar"). How the costs of changing expectations are distributed among the citizens depends not least on their states' choice of the strategy of transformation and its implementation (see "Why the Arguments Have Different Normative Implications"). States with well above-average emissions levels have had, for some time, a duty to reduce emissions levels, and these states themselves are able to take authoritative and effective measures to lower emissions levels in such a way that the reduction is as compatible as possible with the continuation of the lifestyles and projects of their residents. From the perspective of distributive justice, inequalities in these costs should be fairly balanced if they are unavoidable. This is a requirement of domestic justice in the transition to an almost emission-free economy and society.

This way of understanding the differing normative implications of the consequences of historical emissions means that high-emitting states may not claim a higher share of the remaining permissible global budget of emissions owing to the status quo-expectations of their citizens and the special costs they may have in reducing their emissions. Rather, the two justified ways of taking into account historical emissions will reduce what can be considered the fair shares allotted to high-emitting countries. Arguably, an international agreement obliging high-emitting states to stay within their fair shares when understood this way is presently infeasible. However, given that taking into account historical responsibility in the two ways specified is well justified and is supported as a matter of international fairness, one should adopt a diachronically dynamic view of feasibility (see "Feasibility Concerns"). It is possible to change (so-called "soft") feasibility constraints and some of the constraints that restrict the ability of high emitters to reduce emissions.[28] High-emitting states should credibly commit themselves to changing these constraints and thus to a transformational process that will allow them to give increasingly more weight to the justified considerations of historical responsibility that apply to them. From the perspective of developing countries, this is not only a matter of justice—any morally acceptable international agreement will require all states to contribute to staying within the intergenerationally fair limits of a global emissions budget. But also, from the perspective of developing states, this is a matter of feasibility: any agreement that requires their national emissions budgets be unnecessarily and unfairly reduced because of an unfair unwillingness of high-emitting states to become historically responsible agents is an unacceptable agreement. At the very least, it is difficult to see how governments of developing countries could publicly legitimize becoming parties to such an agreement otherwise.[29]

NOTES

1. See Williges et al. (under review) for a discussion of what can be considered a fair allocation and what budgets result from it.

2. Individuals are entitled to a share of the remaining global budget, or so is the assumption in this article. Further, it is assumed that all people are entitled to equal consideration (for a justification see Meyer 2013). Accordingly, the fair allocation of the remaining emissions budget among states is based on an understanding of the fair distribution among people. If differences in entitlement between individuals cannot be justified, each individual ought to be allocated an equal share of the remaining global budget. If so, the EPC principle will determine the legitimate respective shares of the states by the number of their current and future residents only.

3. To be sure, how many benefits per emission (attributed to them) states will in fact realize, in part will depend on their transformation strategies and their success in implementing them. Since the allocation of the limited GCB is done for the whole transformation period, the allocation does encourage each state to make the most efficient use of its national emissions budget (also by selling emissions rights if doing so will maximize the state's benefits). Some states may already have engaged in policies of reducing emission-intensity and successfully so. Thus, they realize the same amount of benefits as others, but with less emissions. States' successes achieved in reducing emission-intensity should benefit them. However, as I argue that we should

allocate the GCB from 1990 onwards (since at least from 1990 states are liable for the emissions caused) these successes will benefit them. Here I assume that states have not begun to engage in reducing emission-intensity before 1990 or that only efforts since 1990 are normatively significant.

4. This is a rather rough and somewhat dated estimate. To clarify this information, I have provided the share of worldwide emissions for several countries and groups of countries from 1850 to 2014 (sum of national emissions / sum of global emissions) based on Gütschow et al. (2016). The shares differ depending on whether one takes into account land use, land-use change, and forestry (LULUCF), that is, emissions and removals of greenhouse gases resulting from direct human-induced land use such as settlements and commercial uses, land-use change, and forestry activities. The first number includes LULUCF, but the second excludes it. The country share of emissions for the United States is 22.2 percent and 27.3 percent, for the EU countries (EU-28) 17.1 percent and 23.3 percent, for the least developed countries (LDCs) 3.0 percent and 0.4 percent, and for the BASIC countries (Brazil, South Africa, India, China) 22.1 percent and 17.2 percent. On the basis of these data the United States and the EU-28 countries - together caused thirteen times (incl. LULUCF) or 126 times (excl. LULUCF) as many emissions as the LDCs from 1850 to 2014. Thanks to Keith Williges for the calculations.

5. For the purposes of this chapter, I am bracketing the equally important and controversial question of who has to provide measures of compensation for unavoidable climate change and what must be considered unavoidable climate damages (but see Meyer 2013, 609–14; Wallimann-Helmer et al. 2019). The answer to this question may inform how emissions rights are allocated going forward.

6. To be precise (see Gütschow et al. 2016) from 1850 to 1975: 259 gigatonnes; 1975–2012: 206 gigatonnes. Thanks to Keith Williges.

7. The nonidentity problem is based on the assumption that future people cannot be harmed (or benefited) by actions that are among the necessary conditions of their existence as individuals (see Parfit 1984, 351–79). Here it is assumed that this is correct. But elsewhere (Meyer 2003) I argued, first, this assumption presupposes a certain understanding of harm-doing: people are harmed or benefited by an action, if the action made them worse or better off than they would have been had the action been omitted. Second, the alternative threshold understanding of harm can evade the nonidentity problem: people are harmed by actions if they fall below a certain threshold of well-being. It is not a question of comparing how they would have fared had the action been omitted. Accordingly, such actions can be identified as harmful to future people, even if these actions also are constitutive of their existence.

8. See Working Group I (1990, 5–6). This is a very simplistic assumption. For a discussion of when people can be ascribed knowledge of the consequences of climate change and their possible actual ignorance can be blameworthy, see for example Gosseries (2004, 36, 40) (list and discussion of some data that could be seen as an alternative to 1990: 1840 (as suggested in the Brazilian Proposal), 1896 (first scientific text on the greenhouse effect by Sante Arrhenius), 1967 (first serious modeling attempts), and 1995 (the second IPCC Report)). The approach of aiming at setting a certain point in time from which on all actors can be assumed equally to have had

access to the relevant knowledge base or from which on they equally are liable for being ignorant is relying on simplistic assumptions, as Gardiner points out (Gardiner 2016, 111–13): Gardiner documents an evolution of awareness over time and argues that individuals and other (especially state) agents have had different access to relevant knowledge about the consequences of their emissions-generating actions and that this is relevant for a differentiated historical and agent-relative attribution of responsibility to avoid risk-impositions. Acknowledging the complexity of the issue, for the purpose of this chapter I will work with the simplified assumption that people and agents have been liable to know about the consequences of their emissions-generating activities since 1990.

9. Here I make the following assumptions: first, it can be shown that without taking historical emissions into account, even nonstrictly egalitarian principles of distribution, in particular the so-called "priority view," require at least the equal allocation of emissions rights if not larger shares for many people living in so-called "developing countries" (Meyer and Roser 2010, 231–33). Second, the EPC approach concerns the allocation of the remaining GCB among countries not persons. Third, states are responsible to distribute the remaining emissions permits in a legitimate way and (in my understanding) in a way that minimally requires that the basic rights of all persons residing in the country are being secured (see Meyer and Pölzler forthcoming). Fourth, given a strict limit of their national budgets, states (and substate political authorities) have strong incentives to make the most efficient use of their budgets.

10. But see fn. 3 above.

11. The total emissions from 1850 to 2014 (including LULUCF) are 2,091 gigatonnes, emissions caused by currently living people since eighteen years of age (including LULUCF) are 753 gigatonnes, and emissions caused by currently living people older than eighteen between 1990 and 2014 (including LULUCF) are 242 gigatonnes. Thus, the share of emissions caused by currently living people older than eighteen between 1990 to 2014 (including LULUCF) in the total emissions 1850 to 2014 is ca. 11 percent. Thanks to Keith Williges for the calculations based on UN Population data (United Nations, Department of Economic and Social Affairs, Population Division (2019)) and Gütschow et al. (2016).

12. In limiting the scope of the argument to post-1990 emissions-generating activities, I acknowledge that some (and apparently Posner and Weisbach 2007, 104–16) object to having to bear the distributive consequences of their actions when they cannot be blamed for them. If so, we will also have to limit the argument further and, in particular, to emissions-generating activities of competent adults. This is reflected in the numbers provided in fn. 9 and the accompanying text where I only take into account emissions by currently living people older than eighteen between 1990 and 2014, and their share in the total emissions.

However, I would like to note that I do not find this line of argument fully convincing. The distributive justice justification of taking into account benefits from emissions-generating activities does not rely on the attribution of blame. The distribution principle on which the argument is based does not only relate to wrongful actions but also to unequal distributions of well-being owing to emissions-generating activities on the assumption that people, wherever they live, have an equal claim to benefits from emissions-generating activities. The argument justifies the idea of equalizing emission benefits over the whole lifetime of individuals. If so, it is important for the distribution principle that an individual has already realized benefits from emissions-generating activities. It does not matter whether this happened knowingly and wrongfully. The issue is not one of providing measures of compensation for misconduct of people from industrialized countries. Of course, the practical relevance of this dispute diminishes over time, as the proportion of emissions caused after 1990 increases and fewer and fewer people are blamelessly ignorant since they have become competent adults. However, as the numbers provided in fn. 11 show, the difference of the share by currently living people since eighteen years of age (including LULUCF) in the total emissions is ca. 35 percent, whereas the share by currently living people older than eighteen between 1990 to 2014 is ca. 11 percent.

13. In this contribution, I do not investigate the relevance of the harmful consequences of historical and past emissions-generating activities in terms of losses and damages that are highly unequally distributed.

14. Total historical emissions from 1850 to 1990 (including LULUCF): 1,328 gigatonnes; emissions embodied in capital stock in 1990: 62.7 gigatonnes. Thus, the ratio of emissions embodied in capital stock in relation to all historical emissions (from 1850 to 1990) is ca. 5 percent. Thanks to Keith Williges for the calculations. For the emissions data, see Gütschow et al. (2016), for the capital stock data see Feenstra et al. (2015) (global capital stock in 1990 as estimated by Feenstra is converted from a monetary value to the embedded carbon content by a multiplier (e.g., capital stock is multiplied by 0.000426357 gigatones carbon per billion USD of capital stock) derived from Müller et al. (2013)).

15. Assuming that the good "benefits from emissions-generating activities" as such makes sense as the subject of an investigation from a distributive point of view. For this debate, see Baatz and Ott (2017).

16. For alternative interpretations of the normative significance of historical emissions and their consequences, see the contributions in Meyer and Sanklecha (2017a).

17. See fns. 9 and 10 above.

18. The net benefits of the consequences of previously living people include the actually realized and expected benefits from historical emissions of currently living people, the actual and expected adaptation costs of currently living people, and the unavoidable damages or losses that they have suffered or are likely to suffer in their lifetime.

19. Again (see fns. 9 and 11 above), total emissions 1850 to 2014 (including LULUCF): 2,091 gigatonnes; emissions caused by currently living people since eighteen years of age (including LULUCF): 753 gigatonnes; emissions embodied in capital stock in 1990: 62.7 gigatonnes. Thus, historical emissions not benefiting those currently alive amounts to 1,275.3 gigatonnes or ca. 61 percent of total emissions.

20. The three objections do not speak against the two proposals to take past emissions into account, because according to these proposals the consequences of historical emissions are distributively relevant and their relevance in no way is based on the idea of compensation for past injustices. The argument in favor of taking these subsets of historical emissions into account is not that the emissions caused wrongful

harm, so that the injured party (or those who wrongfully receive benefits from the wrongful and harmful acts) must provide measures of compensation to those who (indirectly) are harmed. The three objections speak against such an understanding, based on the idea of compensatory justice, of the normative significance of historical emissions and their consequences (see Meyer 2013).

21. The expectation of continuing to cause high emissions is impermissible if causing the requisite amount of emissions is not necessary in order to pursue the project or if the project can be replaced by another less emission-intensive project that enables the actor to realize the same or very similar benefits. For details of the argument, see Meyer and Sanklecha (2011; 2014).

22. This is because the states in which these processes took place are liable for their harmful (long-term) consequences within and outside their national borders (Thompson 2017; Butt 2017).

23. In this and the following section, I present some of the research results from Meyer and Sanklecha (2011; 2014) but reinterpret their significance.

24. In making this claim, I assume that the particular premises of both arguments are plausible and that the arguments are conclusive.

25. Above I have argued that grandfathering on which CAC relies cannot be justified and that taking into account legitimate expectations cannot justify grandfathering in terms of the allocation of GCB among states.

26. Again (see fn. 9 above), emissions from 1990 to 2014 (including LULUCF): 791.3 gigatonnes; emissions from 1850 to today (including LULUCF): 2,091 gigatonnes. Thus, the share of emissions from 1990 to 2014 in total emissions is ca. 38 percent (see Gütschow et al. 2016).

27. Broome (2012, 47–48) argues that for reasons of urgency and feasibility one should opt for what he calls the "efficiency without sacrifice" strategy. At the same time he considers implementing this strategy as a politically prudent first step toward pursuing a strategy that will require sacrifices of the high-emitting countries that Broome believes justice requires (and that can be shown to be equally efficient) (ibid.). However, if the strategy of "efficiency without sacrifice" were implemented, more sacrifices of high-emitting countries would be required to move from the thus realized state of affairs to the just solution. As Brennan and Sayre-McCord (2016, 441) argue, implementing Broome's strategy "efficiency without sacrifice" will make the implementation of a just allocation of the burdens of preventing dangerous climate change less feasible. If so, then pursuing this strategy cannot be considered a politically prudent first step toward a just solution.

28. Consider, for example, that the European Commission in The European Green Deal (2019) has not only committed the EU to "no net emissions of greenhouse gases in 2050" (2) but also promises (4-5): "By summer 2020, the Commission will present an impact assessed plan to increase the EU's greenhouse gas emission reductions target for 2030 to at least 50% and toward 55% compared with 1990 levels in a responsible way. To deliver these additional greenhouse gas emissions reductions, the Commission will, by June 2021, review and propose to revise where necessary, all relevant climate-related policy instruments."

29. The contribution is based on earlier articles, namely: Meyer and Roser (2010) and Meyer (2013) on the relevance of historical emissions for the fair allocation of the remaining global carbon budget and Meyer and Sanklecha (2011; 2014) on the relevance of legitimate expectations for the fair distribution of the costs of the transition to a virtually emission-free society. The respective arguments as developed seem to be in tension or incompatible. This was already the subject of lectures at the Observer Research Foundation, New Delhi (on September 25, 2018, with comments by Shashi Motilal and Sanjay Vashist), at the University of Marburg, Department of Philosophy (October 30, 2018), at Princeton University—organized by Environmental Institute, Climate Futures Initiative, University Center for Human Values (December 4, 2018, together with Pranay Sanklecha and with a comment by Peter Singer) and at the workshop of the Wittgenstein Symposium (August 8, 2019). Participants' very helpful criticism and many constructive comments have contributed to the development of the new theses of this article. I would also particularly like to thank my colleagues in Graz, Gottfried Kirchengast (climate physics), Karl Steininger and Keith Williges (both climate economists), Jeroen Hopster, Thomas Pölzler, and Santiago Truccone (all three in philosophy) for discussion and the coeditor of this volume, Sarah Kenehan, for her many insightful comments and suggestions. Special thanks also to Benedikt Namdar (student assistant in Working Unit Practical Philosophy at the University of Graz).

Special thanks to the editors of this volume, Sarah Kenehan and Corey Katz for their extremely helpful comments and criticisms. All remaining errors are mine.

REFERENCES

Baatz, Christian, and Konrad Ott. 2017. "In Defense of Emissions Egalitarianism?" In *Climate Justice and Historical Emissions*, edited by Lukas H. Meyer and Pranay Sanklecha, 165–97. Cambridge: Cambridge University Press.

Baumert, Kevin, Timothy Herzog, and Jonathan Pershing. 2005. *Navigating the Numbers: Greenhouse Gas Data and International Climate Policy*. Washington, DC: World Resources Institute.

Brennan, Geoffrey and Geoffrey Sayre-McCord. 2016. "Do Normative Facts Matter . . . to What Is Feasible?" *Social Philosophy and Policy* 33 (1–2): 434–56.

Broome, John. 2012. *Climate Matters: Ethics in a Warming World*. New York: Norton.

Butt, Daniel. 2017. "Historical Emissions." In *Climate Justice and Historical Emissions*, edited by Lukas H. Meyer and Pranay Sanklecha, 61–79. Cambridge: Cambridge University Press.

European Commission. 2019. "Communication from the Commission." The European Green Deal, December 11. https://eur-lex.europa.eu/legal-content/EN/TXT/?qid=1588580774040&uri=CELEX:52019DC0640.

Feenstra, Robert C., Robert Inklaar, and Marcel P. Timmer. 2015. "The Next Generation of the Penn World Table." *American Economic Review* 105: 3150–82.

Gardiner, Stephen M. 2016. "In Defense of Climate Ethics." In *Debating Climate Ethics*, edited by Stephen M. Gardiner and David Weisbach, 1–132. New York: Oxford University Press.

Gilabert, Pablo. 2012. "Comparative Assessments of Justice, Political Feasibility, and Ideal Theory." *Ethical Theory and Moral Practice* 15 (1): 39–56.

Gilabert, Pablo, and Holly Lawford-Smith. 2012. "Political Feasibility. A Conceptual Exploration." *Political Studies* 60 (4): 809–25.

Gosseries, Axel. 2004. "Historical Emissions and Free Riding." *Ethical Perspectives* 11 (1): 36–60.

Gütschow, Johannes, M. Louise Jeffery, Robert Gieseke, Ronja Gebell, David Stevens, Mario Krapp, and Marcia Rocha. 2016. "The PRIMAP-hist national historical emissions time series." *Earth System Science Data* 8: 571–603.

Holtung, Nils, and Kasper Lippert-Rasmussen. 2007. "An Introduction to Contemporary Egalitarianism." In *Egalitarianism: New Essays on the Nature and Vale of Equality,* edited by Nils Holtung and Kasper Lippert-Rasmussen, 1–37. Oxford: Oxford University Press.

Hurka, Thomas. 1993. *Perfectionism*. Oxford: Oxford University Press.

Jewell, Jessica, and Aleh Cherp. 2020. "On the political feasibility of climate change mitigation pathways: Is it too late to keep warming below 1.5°C?" *WIREs Climate Change* 11 (1): 1–12.

Kenehan, Sarah. 2017. "In the Name of Political Possibility." In *Climate Justice and Historical Emissions,* edited by Lukas H. Meyer and Pranay Sanklecha, 198–218. Cambridge: Cambridge University Press.

Lawford-Smith, Holly. 2013. "Understanding Political Feasibility." *Journal of Political Philosophy* 21 (3): 243–59.

Meyer, Lukas H. 2003. "Past and Future. The Case for a Threshold Conception of Harm". In *Rights, Culture, and the Law. Themes from the Legal and Political Philosophy of Joseph Raz,* edited by Lukas H. Meyer, Stanley P. Paulson, and Thomas W. Pogge. 143–59. Oxford: Oxford University Press.

Meyer, Lukas H. 2009. "Intergenerationelle Suffizienzgerechtigkeit. ". In *Generationengerechtigkeit: Ordnungsökonomische Konzepte,* edited by Nils Goldschmidt. 281–322. Tübingen: Mohr Siebeck.

Meyer, Lukas H. 2013. "Why Historical Emissions Should Count." *Chicago Journal of International Law* 13 (2): 597–613.

Meyer, Lukas H. 2015. "Die Grundstruktur als institutionelle Ausprägung von John Rawls' Gerechtigkeit als Fairness. " In *John Rawls: Politischer Liberalismus*, edited by Otfried Höffe, 147–62. Berlin: de Gruyter.

Meyer, Lukas H., and Thomas Pölzler. Forthcoming. "Basic Needs and Sufficiency: The Foundations of Intergenerational Justice." In *Oxford Handbook of Intergenerational Ethics*, edited by Stephen Gardiner. Oxford: Oxford University Press.

Meyer, Lukas H., and Dominic Roser. 2006. "Distributive Justice and Climate Change: The Allocation of Emission Rights." *Analyse & Kritik* 28 (2): 223–49.

Meyer, Lukas H., and Dominic Roser. 2010. "Climate Justice and Historical Emissions." *Critical Review of International Social and Political Philosophy* 13 (1): 229–53.

Meyer, Lukas H., and Pranay Sanklecha. 2011. "Individual Expectations and Climate Change." *Analyse & Kritik: Zeitschrift fuer Sozialtheorie* 32 (2), 449–71.
Meyer, Lukas H., and Pranay Sanklecha. 2014. "How Legitimate Expectations Matter in Climate Justice." *Politics, Philosophy & Economics* 13 (4): 369–93.
Meyer, Lukas H., and Pranay Sanklecha, eds. 2017a. *Climate Justice and Historical Emissions*. Cambridge: Cambridge University Press.
Meyer, Lukas H., and Pranay Sanklecha. 2017b. "Introduction: On the Significance of Historical Emissions for Climate Ethics." In *Climate Justice and Historical Emissions*, edited by Lukas H. Meyer and Pranay Sanklecha, 1–21. Cambridge: Cambridge University Press.
Meyer, Lukas H., and Timothy Waligore. 2018. "Die Aufhebungsthese. Grundlinien einer Theorie des gerechten Umgangs mit historischem Unrecht. " In *Zeit—eine normative Ressource?*, edited by Frank Dietrich, Johannes Müller-Salo, and Reinhold Schmücker, 215–30. Frankfurt am Main: Klostermann.
Müller, Daniel B., Gang Liu, Amund N. Løvik, Roja Modaresi, Stefan Pauliuk, Franciska S. Steinhoff, and Helge Brattebø. 2013. "Carbon emissions of infrastructure development." *Environmental Science & Technology* 47 (20): 11739–46.
Ortner, Florian, Thomas Pölzler, Lukas H., Meyer, and Oliver Sass. 2017. "Natural hazards and the normative significance of expectations in protecting alpine communities." In *Geophysical Research Abstracts: Abstracts of the European Geosciences Union General Assembly 2017*, edited by European Geosciences Union. Vienna, Munich: Copernicus.
Parfit, Derek. 1984. *Reasons and Persons*. Oxford: Clarendon Press.
Posner, Eric and David Weisbach. 2007. *Climate Change Justice*. Princeton: Princeton University Press.
Rawls, John. 1999. *A Theory of Justice, Revised ed.* Cambridge: Harvard University Press.
Rawls, John. 2005. *Political Liberalism, Expanded ed.* New York: Columbia University Press.
Thompson, Janna. 2017. "Historical Responsibility and Climate Change." In *Climate Justice and Historical Emissions*, edited by Lukas H. Meyer and Pranay Sanklecha, 46–60. Cambridge: Cambridge University Press.
UNFCCC (United Nations Framework Convention on Climate Change). 2015. "The Paris Agreement." Accessed July 22, 2020. http://unfccc.int/paris_agreement/items/9485.php.
United Nations Population Fund. 2014. "State of World Population." Accessed March 3, 2020. https://www.unfpa.org/swop-2014.
UN (United Nations), Department of Economic and Social Affairs, Population Division. 2019. "World Population Prospects 2019." 1. Accessed March 3, 2020. https://population.un.org/wpp/Download/Standard/Population/.
Wallimann-Helmer, Ivo, Lukas H. Meyer, Kian Mintz-Woo, Thomas Schinko, and Olivia Serdeczny. 2019. "The Ethical Challenges in the Context of Climate Loss and Damage: Concepts, Methods and Policy Options." In *Loss and Damage from Climate Change. Concepts, Methods and Policy Options*, edited by Reinhard

Mechler, Laurens M. Bouwer, Thomas Schinko, Swenja Surminski, and JoAnne Linnerooth-Bayer, 39–62. Berlin, Heidelberg: Springer.

Weisbach, David. 2016. "The Problems with Climate Ethics." In *Debating Climate Ethics*, edited by Stephen M. Gardiner and David Weisbach, 134–243. New York: Oxford University Press.

Williams, Bernard. 1973. "A critique of utilitarianism." In *Utilitarianism For & Against*, edited by Jack Smart and Bernard Williams, 80–150. Cambridge: Cambridge University Press.

Williges, Keith, Lukas H. Meyer, Karl Steininger, and Gottfried Kirchengast. Under Review. "Fairness Critically Conditions the Carbon Budget Allocation Across Countries."

Working Group I to the First Assessment Report of the Intergovernmental Panel on Climate Change. 1990. *Climate Change: The IPCC Scientific Assessment*. Cambridge: Cambridge University Press.

Chapter 6

Toward Climate Justice

Making the Polluters Pay for Loss and Damage

Md Fahad Hossain, Danielle Falzon,
M. Feisal Rahman, and Saleemul Huq

CLIMATE JUSTICE AND LOSS AND DAMAGE

What is the best strategy to feasibly enact climate justice? After nearly three decades of international negotiations and research on climate change, it is now well known that industrialized, developed countries have contributed the most to cause the problem, while developing countries have contributed nearly nothing but will experience the worst effects of climate change (Roberts and Parks 2007). Even worse, developed countries have made much of their wealth from the activities that caused climate change, on top of extracting resources and labor from developing countries, while developing countries have limited funds and resources with which to prepare for the impacts. In the UN climate negotiations, loss and damage due to climate change has emerged as a central concern of developing countries, and one of the key components of this concern is the issue of how to generate funding for countries experiencing the effects of climate change. However, calls for loss and damage finance have bumped up against political, economic, and technical challenges that call into question the feasibility of developing a dedicated loss and damage fund. In what follows, we present a partial solution to the problem of loss and damage finance. We argue that an international fund for loss and damage is possible as long as the money is sourced according to the polluter pays principle. Such an approach can sidestep the feasibility challenges that traditional climate finance structures present, and ensure that climate justice is partially delivered from those causing climate change to those experiencing the effects.

I. A BRIEF INTRODUCTION TO LOSS AND DAMAGE

In contemporary international climate politics, climate justice goes hand-in-hand with the issue of loss and damage. Loss and damage is considered a new issue in international climate policymaking relative to the two more established pillars of mitigation and adaptation. While mitigation works to reduce the greenhouse gas (GHG) emissions and activities causing climate change, and adaptation aims to reduce the impact of climate change on communities around the world, loss and damage accounts for those impacts that do occur. Due to years of insufficient action on climate change, we know that the world will face massive losses and damages even with mitigation and adaptation activities (IPCC 2018). Already some of the world's most vulnerable nations and communities are facing the effects (IPCC 2018; Marino 2015; McAdam 2010; Huq et al. 2019). While we know that something must be done to address loss and damage in the international policymaking arena, it is still unclear what actions would be most effective and appropriate.

One of the reasons why addressing loss and damage is complicated is due to its connection to climate justice. As explained above, countries that will experience the worst effects of climate change are those that contributed to it the least (Working Group II 2007; Roberts and Parks 2007). These are primarily small island developing states (SIDS) and least developed countries (LDCs) that did not reap the benefits of industrialized development as the world's wealthiest countries did. They are therefore also the most likely to experience the greatest losses and damages, and lack the resources to take preventative action on their own. According to the logic of climate justice, wealthy countries should assist poor and vulnerable countries through finance, technology, knowledge transfer, adaptation, and mitigation activities (Kasperson and Kasperson 2001; Robinson 2018; Schlosberg 2012). However, obligations that resemble compensation or liability are still shunned in international policymaking.

In this context, we must ask: How can we ensure that loss and damage solutions contribute to climate justice? This question is complex because we cannot fully anticipate the losses and damages that will occur due to climate change impacts, and therefore cannot necessarily know in advance what is needed. Loss and damage may occur from rapid onset events such as cyclones or flash floods, or from slow-onset events such as sea level rise. Furthermore, while some losses and damages can be calculated based on the economic value of assets lost, many losses and damages are noneconomic—such as loss of lives and damages to mental and physical health—and therefore cannot be quantified in the same way. Ultimately, we argue in this chapter that any solution to loss and damage must begin with the polluters who contributed (and continue to contribute) to climate change. If we start by assuming that fairness demands that the polluter must pay, we can establish the basis for an approach that works toward climate justice.

II. HISTORY OF LOSS AND DAMAGE IN UNITED NATIONS FRAMEWORK CONVENTION ON CLIMATE CHANGE

Pre-Warsaw International Mechanism

Discussion on loss and damage is as old as the climate regime itself. Back in 1991, when the United Nations Framework Convention on Climate Change (UNFCCC) was under negotiation, Vanuatu, on behalf of the newly formed Alliance of Small Island States (AOSIS), put forward a proposal to set up an international insurance pool that would "compensate the most vulnerable small island and low-lying coastal developing countries for loss and damage arising from sea level rise" (INC 1991, 2). The mandatory contributions to the proposed insurance pool would come from the industrialized nations and would be calculated based on a modality drawn from the "Brussels Supplementary Convention on Third Party Liability in the field of Nuclear Energy 1963." The proposal was not accepted. In fact, the term "loss and damage" did not receive any formal recognition.

However, the proposal did have some impact on the Convention. Article 4.8 of the UNFCCC makes a passing remark that action may be necessary "to meet the specific needs and concerns of developing country Parties arising from the adverse effects of climate change" (UNFCCC 1992, 8). Also, article 4.4 notes that developed countries are required to "assist the developing countries that are particularly vulnerable to the adverse effects of climate change in meeting the costs of adaptation to those adverse effects" (ibid.).

Warsaw International Mechanism

More than a decade after the adoption of the UNFCCC, in 2007, the concept of "loss and damage" was incorporated into the international climate regime for the first time at the thirteenth session of the Conference of the Parties (COP13).

At COP 19 in 2013, the Warsaw International Mechanism on Loss and Damage (WIM), to be administered by an Executive Committee (ExCom), was established under the Cancun Adaptation Framework to "address loss and damage associated with impacts of climate change, including extreme events and slow onset events, in developing countries that are particularly vulnerable to the adverse effects of climate change" (UNFCCC 2014, 6). Functions of the WIM include, among others: (1) enhancing knowledge and understanding of risk management approaches to address loss and damage; (2) strengthening dialogue, coordination, coherence, and synergies among relevant stakeholders; and (3) enhancing action and support, including finance, technology, and capacity-building (ibid).

The 2019 review of WIM at COP 25 focused on finance and governance. On finance, developing countries called for "adequate, easily accessible, scaled up, new and additional, predictable finance" (G77 and China 2019, 1) that recalls their demand for a finance mechanism for loss and damage. However, the outcomes did not live up to their expectations. The Board of the Green Climate Fund (GCF) was merely requested to "continue providing financial resources" within its existing structures (UNFCCC 2020, 7). Established in 2010 and launched in 2014, the GCF is the largest operating entity of the financial mechanism under the UNFCCC. Currently, it has two funding windows—adaptation and mitigation (UNFCCC 2012) to support developing countries in climate action-meaning its existing structure does not have a mandate for financing loss and damage. At COP 25, ExCom was likewise asked to "clarify how developing country Parties may access funding" for the purpose of putting together proposals related to the strategic workstreams of its five-year rolling work plan (UNFCCC 2020, 8). They were also asked to establish an expert group to prepare a plan of action on the subject of "finance, sources of support . . . ," by the end of 2020 (Puig, Wewerinke-Singh and Huq 2019, 1). Nevertheless, this progress in finance is "encouraging" (Calliari et al. 2020).

On the second issue of governance, the WIM was established under the Convention, and article 8.2 of the 2015 Paris Agreement states that it "shall be subject to the authority and guidance of the Conference of the Parties serving as the meeting of the Parties to this Agreement" (UNFCCC 2016). This wording led to confusion around whether WIM will continue to be governed under the Convention, under the Paris Agreement, or both. Developed countries, particularly the United States, were in favor of placing it under the Paris Agreement, whereas the developing countries wanted dual governance by both the Convention and the Agreement. COP 25 failed to make a decision on this matter, so discussions will continue at COP 26 (UNFCCC 2020). Even so, one positive outcome of this process was the creation of the Santiago Network "to catalyse the technical assistance of relevant organizations, bodies, networks and experts" in developing countries (UNFCCC 2020, 8), as desired by them (G77 and China 2019). The next review of WIM is scheduled to take place in 2024, and the following ones every five years after.

Article 8 of the Paris Agreement

Loss and damage was further solidified as a third pillar in international climate policymaking through Article 8 of the 2015 Paris Agreement (Okereke and Coventry 2016; Mechler and Schinko 2016; Kreienkamp and Vanhala 2017; Calliari et al. 2019). The article begins with the statement:

> Parties recognize the importance of averting, minimizing and addressing loss and damage associated with the adverse effects of climate change, including extreme weather events and slow onset events, and the role of sustainable development in reducing the risk of loss and damage. (UNFCCC 2015, 12)

The article removes any doubts about the centrality of loss and damage as an issue in climate negotiations and acknowledges the issue's breadth and depth. It furthermore highlights eight action areas: early warning systems; emergency preparedness; slow onset events; events that may involve irreversible and permanent loss and damage; comprehensive risk assessment and management; risk insurance facilities, climate risk pooling and other insurance solutions; noneconomic losses; and resilience of communities, livelihoods, and ecosystems. While many of these areas had already been considered under the WIM, the inclusion of other areas bolstered the argument for why the ExCom should dedicate time and resources to these issues in the future.

While Article 8 was considered a win for many countries and activists who had been working on the issue of loss and damage for years, the negotiations under Article 8 have left much to be desired. Following the entry into force of the Paris Agreement in 2016, Article 8 was conspicuously absent from the agenda of the Ad Hoc Working Group on the Paris Agreement (APA). Instead, loss and damage continued to be negotiated under the Subsidiary Bodies, as an extension of the work of the WIM. It was further left off the agenda during mid-year Subsidiary Body negotiating meetings, a move that prompted critique and frustration at COP 23. Furthermore, the Paris Agreement makes no direct connection between Article 8 on loss and damage and Article 9 on finance, in contrast to direct connections that are made between finance to both mitigation and adaptation. This has created a barrier to arguing for loss and damage finance under the Paris Agreement. It also forecloses the possibility of ensuring compensation or liability for losses and damages (Taub et al. 2016), as we elaborate below.

III. FEASIBILITY IN EXISTING DEBATES OVER LOSS AND DAMAGE

Approaching loss and damage has been a sticky issue in international climate negotiations and in countries' efforts to find workable solutions. Numerous political issues stand in the way of generating a mutually agreeable strategy. Four of these political issues continually arise in discussions, and it remains unclear whether and how they should be considered in approaches to loss and damage. In particular, concerns about the role of insurance, identifying noneconomic losses and damages, attributing events to climate change, and assigning liability and compensation have served as convenient distractions to make action on loss and damage politically, economically, and/or technically infeasible. In order for action to be politically feasible, the literature suggests that actors must accept it as normatively desirable, implement it institutionally, and make political reforms (Lawford-Smith 2013; Gilabert and Lawford-Smith 2012). Economic feasibility requires the funds and funding mechanisms to carry out the action. Technical feasibility means that the actions are logistically possible with the tools and knowledge available. On the issue of loss and damage, insurance, noneconomic loss and damage, attribution, and liability have become roadblocks to progress in the international arena, where disagreements between states ultimately lead to inaction. It is precisely these feasibility conflicts that the proposal we elaborate later in the paper aims to avoid.

Insurance

Even before loss and damage formally entered the UNFCCC negotiating texts, insurance was raised as a potential tool for addressing the impacts of climate change. The use of insurance in disaster relief has served as a workable model, and there are precedents in regional insurance schemes. For example, the Caribbean Catastrophe Risk Insurance Facility was formed in 2007 to fill gaps in existing insurance structures. Insurance has been an active part of the conversation around loss and damage in the past decade. Representatives from insurance organizations have participated in the WIM's Technical Expert Group on Comprehensive Risk Management (TEG-CRM), they have made written submissions to the ExCom, and they attended and spoke at the Suva Expert Dialogue on Loss and Damage in 2018. It is clear that insurance is not only being raised by country delegates as a solution, especially for economic losses such as property damage, but also that insurance companies are interested in working on this issue.

However, others have argued that insurance is at best a partial solution and that there are many barriers to making insurance work for the world's poorest and most vulnerable. Because the existing structures for insurance schemes are in many ways inappropriate to cover the impacts of climate change, the technical feasibility of insurance as an approach to loss and damage makes its utility questionable. First, there is the question of premiums and who can and should pay those premiums. Representatives of vulnerable countries have argued that these costs should not be borne by those nations that did not contribute to climate change, and further cannot be paid by the poor and vulnerable sectors of their populations. Second, insurance will likely not be able to meet the needs of an entire country impacted by a sudden-onset disaster, or if a slow-onset disaster means the losses and damages for a population are constant and continuous. Third, insurance cannot work to address noneconomic losses and damages that are not easily assigned a monetary value. While climate insurance is then part of the action needed to approach loss and damage, it is insufficient to cover the full scope of the problem.

Noneconomic Loss and Damage

Noneconomic loss and damage (NELD) is an especially difficult issue in the conversation on loss and damage. Noneconomic losses are hard to define, quantify, and calculate, and therefore developing actions to address them can present a problem of technical feasibility. Without existing mechanisms for alleviating this sector of losses and damages, they demand new tools. NELD can also create barriers in terms of political feasibility in that different countries and cultures may value noneconomic aspects of their societies differently, leading to disagreements about the normative case for addressing their losses and damages.

NELD can be material or nonmaterial and can have intrinsic or instrumental value (Serdeczny, Waters and Chan 2016). These losses include but are not limited to human life, biodiversity, cultural sites, physical and mental health, sovereignty, education, security, traditions, local knowledge, and sense of place (Fankhauser et al. 2014; Morrissey and Oliver-Smith 2013; Andrei et al. 2014). NELD often goes unaccounted for because it does not have a large or direct impact on economic activity (Fankhauser et al. 2014). Because these losses and damages are noneconomic, economic solutions such as compensation, insurance, or financial support are both insufficient and inappropriate. Rather, they require a deliberate reckoning with more creative approaches that recognize inherent value in things such as nature, culture, and well-being, which existing international organizations and structures are ill-equipped to handle. Incorporating NELD into an approach to loss and damage, however, is critical. If an island nation were to become uninhabitable, it is not only the land, businesses, and property that are lost but the intangible, noneconomic goods will also be lost. Such losses and the suffering that they contribute to must be accounted for.

Attribution

Attribution presents a problem of technical feasibility in calculation and of political feasibility in developing approaches to address loss and damage at the international level. Attribution is defined as "the process of evaluating the relative contributions of multiple causal factors to a change or event with an assignment of statistical confidence" (Working Group I 2013, 872). In the context of loss and damage associated with climate change impacts there are two levels of attribution. The first level is, attributing individual climatic events or processes to anthropogenic GHG emissions. The second level is, attributing specific loss and damage impacts provided that other drivers intervene as well (Roberts and Pelling 2018). While attributing the changed risk of an extreme event to climate change is possible, concluding that the event would not have happened without climate change may never be possible, and clear-cut answers about the sources of specific losses and damages are unlikely (Parker et al. 2015; James et al. 2019; Huggel et al. 2013). Although attribution studies have established strong links between human activities and climate change at global and regional scales, doing the same for losses and damages in a particular community, ecosystem, or economy lies beyond the scope of physical science (James et al. 2019).

While these scientific limits make attribution technically difficult, the political feasibility of assigning attribution presents a more significant barrier. The motivations in the scientific community for assigning attribution are different from the motivations of climate change negotiators. To scientists, the intention is to inquire into the drivers of change, whereas in the negotiations, attribution resurfaces in the context of responsibility, blame, liability, and compensation (James et al. 2019; Parker et al. 2015). Thus, how Parties present attribution science is shaped by their political interests (James et al. 2019). Developed countries tend to highlight only the weakness of attribution science while developing countries focus only on the strength of attribution evidence (James et al. 2019). Even though attribution science has been evolving rapidly, enhancing the understanding about climate change over time, uncertainties in attribution will always remain (ibid.). Scientific advancements are therefore unlikely to resolve the political barriers associated with attribution.

Liability and Compensation

Liability and compensation have been a barrier in the UN negotiations on climate change from the very beginning. Though notions of climate justice suggest that the nations that have benefited from the activities that have caused climate change owe justice to nations that are least responsible and most vulnerable to the projected impacts (Harlan et al. 2015; Kasperson and Kasperson 2001; Roberts and Parks 2007; Robinson 2018), any obligatory action or payment has been politically infeasible and deliberately left out of UNFCCC texts. Instead, vague and broadly interpretable phrases such as "common but differentiated responsibilities" and "new and additional finance" were included in the original Convention and have persisted over the decades of negotiations (UNFCCC 1992). Even in more recent developments in climate finance, such as the establishment of the GCF, countries are only expected to pledge and contribute, rather than being obligated and held accountable for providing finance. The opposition to liability and compensation language has come particularly from the United States, once the world's greatest emitter of GHG emissions. The United States and other wealthy industrialized countries have refused any phrasing that requires such action over the years, so much so that other countries avoid even suggesting it.

On loss and damage, liability and compensation is an especially tricky issue because loss and damage implies an inherent causality: climate change caused these losses and damages and the wealthy industrialized countries caused climate change. Unlike mitigation and adaptation, for which there is some flexibility in proactively defining where and how action will take place, losses and damages require substantive reactive action that helps communities cope with the direct impacts. As such, the pushback that island nations and LDCs received when they attempted to bring conversations about loss and damage into climate negotiations can be understood as wealthy industrialized countries avoiding and rejecting discussions around compensation and liability. By refusing to incorporate liability into the climate negotiations, opposing countries also create a problem of economic feasibility. When countries cannot be held liable for climate change and its impacts, they also cannot be required to contribute funds that would allow impacted communities to be compensated for their losses and damages. Without funds, action on loss and damage is economically impossible.

IV. FINANCE TO FEASIBLY ADDRESS LOSS AND DAMAGE

Why Finance Is Needed

Finance is an indispensable component of climate action and a key indicator of the global community's commitment to materializing article 3.3 of the UNFCCC (Sharma 2016; Richards and Schalatek 2017). Provision for loss and damage finance is mandated under principles of international laws, as well as within the UNFCCC (Richards and Schalatek 2017). Furthermore, the third mandate of the WIM highlights "enhancing action and support, *including finance*" (UNFCCC 2014). However, the initial Work Plan of the WIM ExCom adopted at COP 20 overlooked finance, focusing mainly on the first two functions: enhancing knowledge and strengthening coordination among the relevant stakeholders (CAN and Bond 2017; Richards and Schalatek

2017). The 2018 Suva Expert Dialogue on Loss and Damage also failed to make visible progress with regard to loss and damage finance (Hirsch 2019).

Although the inclusion of loss and damage as a stand-alone article in the Paris Agreement is considered a significant achievement for the Non-annex I parties (Gewirtzman et al. 2018), financing it remains largely unresolved (Taub et al. 2016). Without a financial mechanism, the article offers little more than an endorsement to vulnerable countries that have been pushing for the issue for years (Gewirtzman et al. 2018). As a distinct pillar in the climate regime, loss and damage finance should be separate from and additional to adaptation finance (Climate Focus 2015; Richards and Schalatek 2017).

It is critical that mechanisms for generating loss and damage finance are created sooner rather than later. First, countries and communities are already experiencing losses and damages due to climate change. Increasing heat, more severe and unpredictable weather events, changing weather patterns, and rising seas are currently impacting people's lives and livelihoods around the world. Ensuring that there are funds that can, at the very least, alleviate some of these communities' financial burdens is critical to preventing further losses and damages. Second, mobilizing climate finance has been notoriously difficult, even for mitigation and adaptation funds. Beginning to generate funds to address loss and damage as soon as possible increases the likelihood of having sufficient funding to cover loss and damage as it arises. The experience of existing multilateral climate funds tells us that countries are unlikely to contribute large sums of money and that the bureaucratic challenges may prevent funds from being quickly allocated and distributed. A dedicated funding stream for loss and damage will hopefully protect against these shortcomings and can use more creative means for mobilization.

Challenges to Generating Funds

Although the intention to generate additional funds for loss and damage is evident from WIM's mandate, subsequent negotiations have moved the focus away to market-based, private sector financial instruments resulting in a move away from the solidarity-based proposals, including public sector interventions, and taxation and transfers from developed countries to vulnerable countries (Gewirtzman et al. 2018). As noted by Schafer and Kunzel (2019), funding available in relation to loss and damage is largely inadequate, and developed nations consider financing only some measures for addressing loss and damage. The meager funding is limited primarily to ex-ante risk-reducing measures with little to no available funding for ex-post measures. Funding for both ex-ante and ex-post measures for addressing noneconomic losses is minimal or absent altogether (ibid.).

The reason developed countries tend to shy away from discussions related to financing loss and damage is similar to why they avoided conversations about adaptation finance in the first decade of the UNFCCC regime (Ciplet, Roberts, and Khan 2015). In particular, they feared that an emphasis on adaptation funding might imply an acknowledgment of their responsibility and liability, as their historical GHG emissions have caused climate change. They have the same worry when it comes to the issue of loss and damage: agreeing to finance loss and damage could potentially lead to holding them liable, thereby prompting widespread litigation claiming compensation (Hirsch 2019).

Again, like the case of adaptation (Ciplet, Roberts, and Khan 2015), developed countries view loss and damage as a local issue as opposed to mitigation, which is a global public good (Roberts et al. 2017). Mitigation projects anywhere in the world benefit developing countries and developed countries alike, whereas the benefits related to loss and damage projects will likely be localized.

Furthermore, despite the varying estimates of loss and damage finance needs, it is clear that the gap between what is needed and what is being provided internationally is substantial (Richards and Schalatek 2017; Gewirtzman et al. 2018). Considering that significant gaps already exist in development, humanitarian, and disaster risk reduction finance (all of which are voluntary), as well as in adaptation finance, which involves internationally agreed-upon obligation, generating voluntary funds for loss and damage does not seem realistic (Richards and Schalatek 2017).

V. MAKING THE POLLUTERS PAY FOR LOSS AND DAMAGE

The key issue for discussion and debate around loss and damage is financial support for the victims of climate impacts. While developed countries have primarily promoted insurance as a measure to protect against climatic risks, they remain unwilling to consider other solutions. To be sure, insurance is a key component (indeed, there are several pilot programs going on around the world), but it is no panacea. It is particularly unsuitable for the poorest and most vulnerable communities who cannot afford to pay the insurance premiums, even if they are subsidized.

The polluter pays principle, also part of the broader guideline for sustainable development agreed under the 1992 Rio Declaration, suggests that producers of pollution should bear the costs of managing it (UN 1992). Given the urgency of the climate crisis, the time has come to think about raising money to compensate victims of climate change through innovative sources, applying the polluter pays principle wherever possible. A global fund does not require developed countries to accept liability, since it could be based on solidarity contributions or a tax levied on polluters, but still creates a mechanism for compensatory justice. Instead of waiting for voluntary contributions that are proving to be politically unfeasible, this strategy bypasses the political barriers by taxing the polluters at the source, and distributing the funds to vulnerable nations through an international finance mechanism. Previous reports (Durand et al. 2016, Roberts et al. 2017) have presented several ideas exploring the polluter pays principle. Here we share two such proposals that could be used to raise funds for loss and damage.

Taxing the Polluting Companies

Recent studies quantifying contributions of CO_2 and CH_4 emissions have traced global atmospheric CO_2, surface temperature, sea level rise, and acidification of oceans to major fossil fuel companies and cement manufacturers (Licker et al. 2019; Ekwurzel et al. 2017; Griffin 2017). Griffin (2017) reported that a total of one-hundred extant fossil fuel producers were responsible for 71 percent and 52 percent of global industrial GHG emissions from 1988–2015 to 1854–2015, respectively. The report further identified that 51 percent of the global industrial GHG emissions since 1988 can be traced to just twenty-five corporate and state-owned entities, suggesting that the historical emissions of these polluting entities are large enough to have contributed significantly to global climate change. The fossil fuel industry has also enjoyed government subsidies and regulatory preferences. According to the polluter pays principle, these companies should pay the price for the loss and damage associated with climate change and suffered by those who did not cause the problem.

Recently, the public has learned that not only have these companies knowingly contributed to climate change even after the dangers of doing so became common knowledge, but many of them also knew about the impacts of their activities long before the Intergovernmental Panel on Climate Change's (IPCC) first assessment report in 1990. They have likewise worked systematically to prevent political actions aimed at curbing GHG emissions (Frumhoff, Heede, and Oreskes 2015). A study by *InsideClimate News* in 2015 revealed that Exxon (now ExxonMobil) knew about climate change trends as early as 1977, and still launched campaigns to actively sow doubt about the science to discourage political action (Banerjee, Song, and Hasemyer 2015). While Exxon's actions are particularly outrageous, any fossil fuel company that continues to pollute, and thus contribute to the dangerous consequences of global climate change, must be held financially liable for the impacts of global climate change.

Polluting fossil fuel companies could thus be charged a loss and damage tax, especially on their profits in every jurisdiction they operate. This could be done under the auspices of the UNFCCC and the money collected could be used to create a loss and damage fund to compensate victims of climate change. Since this approach taxes companies for their pollution, it bypasses the need for the state to make a voluntary contribution to finance loss and damage, even in those countries where fossil fuel companies are state-owned. However, the suggested taxes cannot be considered an offset for continued pollution; rather, the ultimate goal has to be ensuring a transition to renewable energy and environmentally friendly ventures. Furthermore, it has to be ensured that the tax is not eventually passed onto the consumers, which would defeat the purpose of the tax and may exacerbate issues of justice (e.g., the poor individuals/nations will have to make sacrifices in their use of fossil fuel, whereas the wealthy individuals/nations may be able to continue business as usual because they can afford the increases).

A Levy on Air Travel

Another proposed innovative tool considers levying airplane tickets to contribute to a global fund to compensate loss and damage victims (Roberts et al. 2017; Durand et al. 2016). This is suggested based on the International Air Passenger Levy (IAPAL) proposal submitted to the UNFCCC by the LDCs group in 2008, which proposes a new purchase tax on international air tickets to raise funds to support investment in adaptation to climate change (Muller 2008). The IAPAL proposal indicated that a levy of 6USD per international air passenger traveling in economy class and 62USD per passenger traveling in business or first class could accumulate up to 8–10 billion USD per year, with minimal administrative costs, to deliver finance targeted specifically to adaptation (Chambwera et al. 2012).

The advantages of taxing air travel passengers for climate losses and damages include directly linking the polluting activity (i.e., air travel) to loss and damage. The financing tool also does not target passengers based on the historic responsibility of their countries (Muller 2008), but it enables passengers from poor countries who have the means to fly to compensate their poorer counterparts who might be the victims of climate change. The modality, as argued by Muller (2008, 1), could provide "an equitable mechanism for victim compensation in a sector which—due to the nonnational character of the emissions—eludes the traditional interpretation of common but differentiated responsibilities in terms of national (historic) emissions."

Having said that, the proposed levy is aimed at raising finance for loss and damage, but not at reducing flight numbers or affecting passenger behavior.

Therefore, it would not mitigate the effects of climate change as it does not reduce the emissions caused by air travel. However, such a mandatory levy would be an improvement on existing voluntary payments for carbon offsets, which do not guarantee that a passenger will contribute even if they are willing to pay (Muller 2008).

Given the devastating impact of the ongoing COVID-19 pandemic on the airline industry at the time of writing, a proposal to tax passengers might not presently be appreciated by the industry or their governments. However, research conducted previously for the IAPAL proposal indicated that the small levy on individual tickets will have minimal to no impact on tourism-dependent economies (Muller 2008; Chambwera et al. 2012). On the other hand, the raised funds offer opportunities for providing compensatory justice, as they will have a significant impact on averting, minimizing, and addressing loss and damage of the poorest and most vulnerable countries and communities. The LDCs group could consider reviving the old IAPAL proposal as the International Air Passenger Loss and Damage Levy (IAPLDL).

VI. THE FEASIBILITY OF THE PROPOSED APPROACHES

The proposed approaches have great potential in terms of their political, economic, and technical feasibility. Politically, these approaches avoid the political tensions that have plagued policymaking on loss and damage throughout its history by bypassing the politics of the state. Though there may be a concern that the cost of new taxes will be passed onto consumers, thereby impacting the national economy and potentially limiting the plan's political and economic feasibility, it avoids notions of national liability and reparations owed to vulnerable countries. Passing the cost onto consumers should also encourage them to move away from these polluting industries and can inspire innovation to transition toward cleaner technologies. In the long run, such a transition would reduce contributions to a loss and damage fund, but it should also reduce losses and damages by mitigating climate change.

Taxing specific industries or introducing a carbon tax can also lead to leakage (when an emission-reduction policy such as a carbon tax inadvertently causes an increase in emissions in other jurisdictions) and distributional effects (the cost of the carbon tax is distributed in an uneven or unfair way) (Conway et al. 2017). Fortunately, there is a range of measures to minimize leakage and distributional effects. For example, multiple tax rates can be set by jurisdictions with provisions for lower rates to sectors deemed at risk of leakage (Conway et al. 2017). Governments should also regularly measure and report effects of a taxing scheme once such a scheme is introduced. Such reports should also include information on the intended and actual usage of collected revenue, which would increase the visibility of the effects of the tax and enhance the transparency and accountability, and thereby increase trust in and credibility of such taxing schemes (Carattini et al. 2017).

There are also important precedents that support the technical feasibility of the proposed approaches. For example, the International Oil Pollution Compensation Fund (IOPCF) gets its contributions from all international oil tanker companies, and in the event of an oil spill, it makes payments to any affected coastal community, regardless of which ship caused the spill. Instead of putting liability on a single company, the program applies the principle of compensation based on collective liability. Similarly, a levy on fossil fuel companies acknowledges the collective liability that these companies' products have on the climate. By compiling the tax revenues into a fund that can be distributed for future losses and damages, no one company has to take responsibility for a specific event or impact. Rather, the companies pay for their pollution generally, and those who suffer the effects of that pollution have access to compensation.

The French government's 2006 Solidarity Levy, on the other hand, demonstrates the technical feasibility of taxing individuals for air travel whereby a levy is imposed on passengers departing from airports situated in French territory. Beyond France, another seven countries—Chile, Congo, Ivory Coast, Madagascar, Mauritius, Niger, and South Korea—have participated in the program since April 2007. The total proceeds generated from France was about EUR 180 million in 2007 and another EUR 22 million from the seven other participating countries (de Ferranti et al. 2008, 92–93). Each country individually decides the amount of their own levy and then collectively agrees to allocate the total funds to support a common cause. For example, the Solidarity Levy supports UNITAID, an international drug purchase facility that combats malaria, tuberculosis, and HIV/AIDS in developing countries (Roberts et al. 2017).

Needless to say, despite these precedents, implementation of similar schemes for financing loss and damage would need careful thinking and in-depth strategic planning. Making such schemes more politically acceptable is thus a key prerequisite for more stringent and effective climate action (Carattini et al. 2017).

VII. CLIMATE JUSTICE IS A BROADER ISSUE (WE ADDRESS A SLICE)

We believe the polluter pays mechanisms that we have highlighted are advantageous, but they are first steps and not solutions to addressing loss and damage or in delivering climate justice on their own. As we detailed above, the issue of loss and damage is multidimensional and complex, and so demands solutions that extend beyond monetary compensations. We are addressing a slice of the issue with these suggestions, namely the lack of financial resources to fund actions to address loss and damage. Unfortunately, contributions to global climate funds and flows of climate finance that are new and additional to traditional official development assistance are not fulfilling even the world's mitigation and adaptation needs. Many wealthy industrialized countries, in particular the United States, have demonstrated that they will not follow through on their promises to the poorest and most vulnerable. By connecting loss and damage finance directly to the pollution causing climate change (and not to governments), these options ensure that polluters pay and that the climate vulnerable receive some form of justice.

A potential shortcoming of our proposal is that, while it holds *individual polluters* accountable for their pollution, it does little to address the historical responsibility of wealthy industrialized countries for causing the problem, and in doing so arguably depoliticizes it. This is not our intention. We believe that the depoliticization of climate change is a fundamentally flawed approach to generating solutions. As Paprocki (2015) argues, failing to account for the conditions that made particular countries and communities vulnerable fundamentally re-entrenches power imbalances and does little to break down the problem's roots. This "anti-politics of climate change" perpetuates the status quo and side-steps climate justice. The measures we have put forward do not directly challenge the power of countries that profited from causing climate change by demanding a redistribution of either the wealth or the costs that are currently unevenly allocated, nor do they account for the colonial and imperial exploitations that created the massive and unjust wealth and resource gaps that exist in the world. Acknowledging the partiality of the solutions offered here is therefore crucial to ensuring that the fight for climate justice and the work to address loss and damage continues.

Despite these shortcomings, we would like to suggest that the steps we have highlighted are indeed political. The effects of climate change disproportionately impact the poor not only between countries but also within countries. Climate change itself was created and continues to be exacerbated by the actions of the wealthy. A tax on major polluting companies and an airline levy each ensure that the wealthy are paying for the damage they cause. These steps avoid the lack of action that we are currently seeing in international climate politics and multilateral climate finance institutions, and instead transcend the limitations of borders by demanding that elites everywhere pay their fair share. It is of course still imperative that we hold nations accountable and that we keep politics at the forefront of our minds in working to address loss and damage, in both mobilizing and distributing funds. However, money is needed right now and these are clear options to ensure that we can begin to get some help to those who need it.

VIII. CONCLUSION

In this chapter, we discuss the feasibility of generating loss and damage finance to serve some form of compensatory climate justice. Underscoring the need for funds to address loss and damage, we have discussed the political, economic, and technical feasibility challenges. We have shared two strategies in line with the polluter pays principle to establish an international fund to support victims of climate impacts: (1) taxing the polluting companies that have knowingly contributed to climate change since the impacts of GHG emissions became common knowledge, and (2) putting a levy on international air tickets that directly links the payment to the polluting activity. We argue that, by connecting loss and damage finance directly to the pollution that has and is continuing to cause climate change, these options could be important first steps toward ensuring climate justice for the vulnerable. The measures we propose are politically feasible in that they do not directly demand a global redistribution of wealth to pay for loss and damage, and technically and economically feasible with related precedents. Our proposal begins to address the issue of climate justice and provides a starting point for further research and discussions.

REFERENCES

Andrei, Stephanie, Golam Rabbani, and Hafij Islam Khan. 2014. "Non-Economic Loss and Damage Caused by Climatic Stressors in Selected Coastal Districts of Bangladesh." Dhaka, Bangladesh: Bangladesh Centre for Advance Studies (BCAS).

Banerjee, Neela, Lisa Song, and David Hasemyer. 2015. "Exxon's Own Research Confirmed Fossil Fuels' Role in Global Warming Decades Ago." *InsideClimate News*. September 16. https://insideclimatenews.org/news/15092015/Exxons-own-research-confirmed-fossil-fuels-role-in-global-warming

Calliari, Elisa, Swenja Surminski, and Jaroslav Mysiak. 2019. "The Politics of (and Behind) the UNFCCC's Loss and Damage Mechanism." In *Loss and Damage from Climate Change: Concepts, Methods and Policy Options*, edited by Reinhard Mechler, Laurens M. Bouwer, Thomas Schinko, Swenja Surminski, and JoAnne Linnerooth-Bayer, 155–78. Cham, Switzerland: Springer.

Calliari, Elisa, Lisa Vanhala, Linnea Nordlander, Daniel Puig, Fatemeh Bakhtiari, Md Fahad Hossain, Saleemul Huq, and Mohammad Feisal Rahman. 2021. "Article 8: Loss and Damage." In *Paris Agreement: A Commentary*, edited by Geert van Calster and Leonie Reins. Cheltenham: Edward Elgar Publishing.

CAN, and Bond. 2017. "Joint Submission on the Strategic Workstream on Loss and Damage Action and Support." Available at: http://www.climatenetwork.org/sites/default/files/can_bond_joint_submission_on_the_strategic_workstream_on_loss_and_damage_finance.pdf

Carattini, S., Carvalho, M. and Fankhauser, S., 2017. How to make carbon taxes more acceptable. London: Grantham Research Institute on Climate Change and the Environment, and Centre for Climate Change Economics and Policy, London School of Economics and Political Science.

Chambwera, Muyeye, Evans Davie Njewa, and Denise Loga. 2012. "The International Air Passenger Adaptation Levy: Opportunity or Risk for Least Developed Countries." London: LDC paper series, IIED.

Ciplet, David, J Timmons Roberts, and Mizan R. Khan. 2015. *Power in a Warming World: The New Global Politics of Climate Change and the Remaking of Environmental Inequality*. Cambridge: The MIT Press.

Climate Focus. 2015. "Loss and Damage in the Paris Agreement (Climate Focus Client Brief on the Paris Agreement IV)." Available at: https://climatefocus.com/sites/default/files/20160214%20Loss%20and%20Damage%20Paris_FIN.pdf

Conway, D., Richards, K., Richards, S., Keenlyside, P., Mikolajczyk, S., Streck, C., Ross, J., Anthony Liu, A. and Tran, A., 2017. *Carbon Tax Guide: A Handbook for Policy Makers* (No. 129668, pp. 1–172). Washington, DC: The World Bank.

de Ferranti, David, Charles Griffin, Maria-Luisa Escobar, Amanda Glassman, and Gina Lagomarsino. 2008. "Innovative Financing for Global Health: Tools for Analyzing the Options." *Brookings Global Health Financing Initiative Working Paper*, 2. https://www.brookings.edu/wp-content/uploads/2016/06/08_global_health_de_ferranti.pdf

Durand, Alexis, Victoria Hoffmeister, J. Timmons Roberts, Jonathan Gewirtzman, Sujay Natson, Romain Weikmans, and Saleemul Huq. 2016. "Financing Options for Loss and Damage: A Review and Roadmap." USA and Bangladesh: Climate and Development Lab (CDL), Brown University and International Centre for Climate Change and Development (ICCCAD).

Ekwurzel, B., J. Boneham, M. W. Dalton, R. Heede, R. J. Mera, M. R. Allen, and P. C. Frumhoff. 2017. "The Rise in Global Atmospheric CO_2, Surface Temperature, and Sea Level from Emissions Traced to Major Carbon Producers." *Climatic Change* 144(4): 579–90.

Fankhauser, Sam, Simon Dietz, and Phillip Gradwell. 2014. "Non-Economic Losses in the Context of the UNFCCC Work Programme on Loss and Damage." Centre for Climate Change Economics and Policy—Grantham Research Institute on Climate Change and the Environment.

Frumhoff, Peter C., Richard Heede, and Naomi Oreskes. 2015. "The Climate Responsibilities of Industrial Carbon Producers." *Climatic Change* 132(2): 157–71.

G77 and China. 2019. "Group of 77 and China Submission on the Review of the WIM and the Report of the WIM Executive Committee." (CRP.SBSTA.i4_SBI.9). Available at: https://unfccc.int/sites/default/files/resource/CRP.SBSTA_.i4_SBI.9.pdf

Gewirtzman, Jonathan, Sujay Natson, Julie-Anne Richards, Victoria Hoffmeister, Alexis Durand, Romain Weikmans, Saleemul Huq, and J. Timmons Roberts. 2018. "Financing Loss and Damage: Reviewing Options under the Warsaw International Mechanism." *Climate Policy* 18(8): 1076–86.

Gilabert, Pablo, and Holly Lawford-Smith. 2012. "Political Feasibility: A Conceptual Exploration." *Political Studies* 60(4): 809–25.

Griffin, Paul. 2017. "The Carbon Majors Database: CDP Carbon Majors Report 2017." London: CDP Worldwide. https://6fefcbb86e61af1b2fc4-c70d8ead6ce d550b4d987d7c03fcdd1d.ssl.cf3.rackcdn.com/cms/reports/documents/000/002 /327/original/Carbon-Majors-Report-2017.pdf?1501833772

Harlan, Sharon L., David Pellow, and Roberts, J. Timmons. 2015. "Climate Justice and Inequality." In *Climate Change and Society: Sociological Perspectives*, edited by Riley E. Dunlap and Robert J. Brulle, 127–63. New York: Oxford University Press.

Hirsch, Thomas. 2019. "Climate Finance for Addressing Loss and Damage: How to Mobilize Support for Developing Countries to Tackle Loss and Damage." (Analysis 91). Berlin, Germany: Brot für die Welt.

Huggel, Christian, Dáithí Stone, Maximilian Auffhammer, and Gerrit Hansen. 2013. "Loss and Damage Attribution." *Nature Climate Change* 3(8): 694–96.

Huq, Saleemul, Jeffrey Chow, Adrian Fenton, Clare Stott, Julia Taub, and Helena Wright, eds. 2019. *Confronting Climate Change in Bangladesh*. New York: Springer Berlin Heidelberg.

INC. 1991. "Vanuatu: Draft Annex Relating to Article 23 (Insurance) for Inclusion in the Revised Single Text on Elements Relating to Mechanisms" (A/AC.237/ WG.II/Misc.13) Submitted by the Co-chairmen of Working Group II (A/AC.237/ WG.II/CRP.8). Available at: https://unfccc.int/sites/default/files/resource/docs/a/ wg2crp08.pdf

IPCC (Intergovernmental Panel on Climate Change). 2018. *Global Warming of 1.5 °C. An IPCC Special Report on the Impacts of Global Warming of 1.5°C Above Pre-industrial Levels and Related Global Greenhouse Gas Emission Pathways, in the Context of Strengthening the Global Response to the Threat of Climate Change, Sustainable Development, and Efforts to Eradicate Poverty*, edited by Valérie Masson-Delmotte, Panmao Zhai, Hans-Otto Pörtner, Debra Roberts, Jim Skea, Priyadarshi R. Shukla, Anna Pirani et al. In Press.

James, Rachel A., Richard E. Jones, Emily Boyd, Hannah R. Young, Friederike E. L. Otto, Christian Huggel, and Jan S. Fuglestvedt. 2019. "Attribution: How Is It Relevant for Loss and Damage Policy and Practice?" In *Loss and Damage from Climate Change: Concepts, Methods and Policy Options*, edited by Reinhard Mechler, Laurens M. Bouwer, Thomas Schinko, Swenja Surminski, and JoAnne Linnerooth-Bayer, 113–54. Cham: Springer.

Kasperson, Roger E., and Jeanne X. Kasperson. 2001. *Climate Change, Vulnerability, and Social Justice*, edited by Risk and Vulnerability Programme, 1–17. Stockholm: Stockholm Environment Institute. http://www.sei.se/dload/2001/sei-risk.pdf

Kreienkamp, Julia, and Lisa Vanhala. 2017. "Climate Change Loss and Damage." (Policy Brief). London: Global Governance Institute, University College London https://www.ucl.ac.uk/global-governance/news/2017/mar/climate-change-loss-and-damage

Lawford-Smith, Holly. 2013. "Understanding Political Feasibility." *Journal of Political Philosophy* 21(3): 243–59.

Licker, R, B Ekwurzel, S C Doney, S R Cooley, I D Lima, R Heede, and P C Frumhoff. 2019. "Attributing Ocean Acidification to Major Carbon Producers." *Environmental Research Letters* 14(12): 124060.

Marino, Elizabeth K. 2015. *Fierce Climate, Sacred Ground: An Ethnography of Climate Change in Shishmaref, Alaska.* Fairbanks: University of Alaska Press.

McAdam, Jane, ed. 2010. *Climate Change and Displacement: Multidisciplinary Perspectives.* Oxford: Hart.

Mechler, Reinhard, and Thomas Schinko. 2016. "Identifying the Policy Space for Climate Loss and Damage." *Science* 354(6310): 290–92.

Morrissey, James, and Anthony Oliver-Smith. 2013. "Perspectives on Non-Economic Loss and Damage: Understanding Values at Risk from Climate Change." Bonn: United Nations University Institute for Environment and Human Security.

Muller, Benito. 2008. "International Air Passenger Adaptation Levy (IAPAL)." *European Capacity Building Initiative Policy Brief, submitted to UNFCCC AWC-LCA, 12.* https://oxfordclimatepolicy.org/sites/default/files/ecbiBrief-IAPAL13Q%26As.pdf

Okereke, Chukwumerije, and Philip Coventry. 2016. "Climate Justice and the International Regime: Before, During, and After Paris." *Wiley Interdisciplinary Reviews: Climate Change* 7(6): 834–51.

Paprocki, Kasia. 2015. "Anti-Politics of Climate Change: Depoliticisation of Climate Change Undermines the Historic Reasons That Made Bangladesh Vulnerable to It." In *The Bangladesh Paradox*, edited by Aunohita Mojumdar, 54–64. Lalitpur, Nepal: The Southasia Trust.

Parker, Hannah R., Rosalind J. Cornforth, Emily Boyd, Rachel James, Friederike E. L. Otto, and Myles R. Allen. 2015. "Implications of Event Attribution for Loss and Damage Policy." *Weather* 70(9): 268–73.

Puig, Daniel, Margaretha Wewerinke-Singh, and Saleemul Huq. 2019. "Loss and Damage in COP25." Available at: https://unepdtu.org/wp-content/uploads/2019/12/ld-cop25.pdf

Richards, Julie-Anne, and Liane Schalatek. 2017. "Financing Loss and Damage: A Look at Governance and Implementation Options." (Discussion paper). Washington D.C.: Heinrich Böll Stiftung North America.

Roberts, J. Timmons, and Bradley C Parks. 2007. *A Climate of Injustice: Global Inequality, North-South Politics, and Climate Policy.* Cambridge, MA: MIT Press.

Roberts, J. Timmons, Sujay Natson, Victoria Hoffmeister, Alexis Durand, Romain Weikmans, Jonathan Gewirtzman, and Saleemul Huq. 2017. "How Will We Pay for Loss and Damage?" *Ethics, Policy & Environment* 20(2): 208–26.

Roberts, Erin, and Mark Pelling. 2018. "Climate Change-Related Loss and Damage: Translating the Global Policy Agenda for National Policy Processes." *Climate and Development* 10(1): 4–17.

Robinson, Mary. 2018. *Climate Justice: Hope, Resilience, and the Fight for a Sustainable Future*. New York: Bloomsbury Publishing.
Schafer, Laura and Vera Kunzel. 2019. "Steps towards Closing the Loss & Damage Finance Gap: Recommendations for COP25." Bonn: Germanwatch e.V.
Schlosberg, David. 2012. "Climate Justice and Capabilities: A Framework for Adaptation Policy." *Ethics & International Affairs* 26(4): 445–61.
Serdeczny, Olivia, Eleanor Waters, and Sander Chan. 2016. "Non-Economic Loss and Damage in the Context of Climate Change: Understanding the Challenges." Discussion Paper/ Deutsches Institut Für Entwicklungspolitik 2016/3. Bonn: Deutsches Institut für Entwicklungspolitik.
Sharma, Anju. 2016. "Precaution and Post-Caution in the Paris Agreement: Adaptation, Loss and Damage and Finance." *Climate Policy* 17(1): 33–47.
Taub, Julia, Naznin Nasir, M. Feisal Rahman, and Saleemul Huq. 2016. "From Paris to Marrakech: Global Politics around Loss and Damage." *India Quarterly: A Journal of International Affairs* 72(4): 317–29.
UN (United Nations). 1992. "Rio Declaration on Environment and Development. Annex I. Report of the United Nations Conference on Environment and Development, 3–14 June." Available at: https://www.un.org/en/development/desa/population/migration/generalassembly/docs/globalcompact/A_CONF.151_26_Vol.I_Declaration.pdf
UNFCCC (United Nations Framework Convention on Climate Change). 1992. "United Nations Framework Convention on Climate Change." Available at: https://unfccc.int/sites/default/files/conveng.pdf
UNFCCC (United Nations Framework Convention on Climate Change). 2012. "Report of the Conference of the Parties on its seventeenth session, held in Durban from 28 November to 11 December 2011. Addendum. Part Two: Action taken by the Conference of the Parties at its seventeenth session (FCCC/CP/2011/9/Add.1). Available at: https://unfccc.int/resource/docs/2011/cop17/eng/09a01.pdf
UNFCCC (United Nations Framework Convention on Climate Change). 2014. "Report of the Conference of the Parties on its nineteenth session, held in Warsaw from 11 to 23 November 2013. Addendum. Part two: Action taken by the Conference of the Parties at its nineteenth session" (FCCC/CP/2013/10/Add.1). Available at: https://unfccc.int/sites/default/files/resource/docs/2013/cop19/eng/10a01.pdf
UNFCCC (United Nations Framework Convention on Climate Change). 2015. "Paris Agreement." Available at: https://unfccc.int/files/essential_background/convention/application/pdf/english_paris_agreement.pdf
UNFCCC (United Nations Framework Convention on Climate Change). 2016. "Report of the Conference of the Parties on its twenty-first session, held in Paris from 30 November to 13 December 2015. Addendum. Part two: Action taken by the Conference of the Parties at its twenty-first session" (FCCC/CP/2015/10/Add.1). Available at: https://unfccc.int/sites/default/files/resource/docs/2015/cop21/eng/10a01.pdf
UNFCCC (United Nations Framework Convention on Climate Change). 2020. "Technologies for Averting, Minimizing and Addressing Loss and Damage in

Coastal Zones." Policy Brief. Executive Committee of the Warsaw International Mechanism for Loss and Damage andTechnology Executive Committee. Available at: https://unfccc.int/sites/default/files/resource/Joint%20Policy%20Brief.pdf.

UNFCCC (United Nations Framework Convention on Climate Change). 2020. "Report of the Conference of the Parties serving as the meeting of the Parties to the Paris Agreement on its second session, held in Madrid from 2 to December 15, 2019. Addendum. Part two: Action taken by the Conference of the Parties serving as the meeting of the Parties to the Paris Agreement at its second session" (FCCC/PA/CMA/2019/6/Add.1). Available at: https://unfccc.int/sites/default/files/resource/cma2019_06a01E.pdf

Working Group I to the Fifth Assessment Report of the Intergovernmental Panel on Climate Change. 2013. *Climate Change 2013: The Physical Science Basis.* Cambridge: Cambridge University Press.

Working Group II to the Fourth Assessment Report of the Intergovernmental Panel on Climate Change. 2007. *Climate Change 2007: Impacts, Adaptation, and Vulnerability.* Cambridge: Cambridge University Press.

Chapter 7

Deficient International Leadership as a Feasibility Constraint

The Case of Multilateral Negotiations on Climate-induced Human Mobility

Jörgen Ödalen and Felicia Wartiainen

Effective leadership is necessary if we are ever going to reach the international agreements needed to successfully respond to climate change. At the same time, the prospects for the emergence of effective leadership are poor. Inefficient or nonexistent leadership on the international level threatens to block any aspirations to realize climate justice.

In response to this situation, it is quite natural that philosophers, given their expertise, urge potential leaders to take unilateral action by arguing that it is their moral duty to do so. For example, Shue (2011, 17) comments that "[h]umanity's so far leaderless approach to dealing with rapidly accelerating climate change embodies a [. . .] profoundly tragic, catch-22 that has, among other twists and contradictions, transmuted justice into paralysis." Shue proceeds to provide philosophical arguments for why nation states *ought* to assume leadership. He believes that what we need is at least one state willing to break the leadership paralysis by unilaterally taking action in the hope that the others will respond to its example. Shue also notes that while this kind of unilateral leadership is not a universally effective tactic, it is a kind of leadership that is "appropriate when one already has a moral duty to act" (Shue 2011, 23–24).

We do not wish to dispute that moral arguments *are* important and that philosophers working on climate ethics are doing an important job. In this chapter, however, we call attention to the fact that the actors who have the moral responsibility to act must do so as leaders in the international community, a tremendously complex activity that requires much more than the support of strong moral arguments.

We understand leadership, in this context, as an asymmetrical relationship of influence where one actor guides or directs the behavior of others toward a certain goal (Underdal 1994, 140). Even if a state is convinced by moral arguments and willing to assume the role of a leader, it might fail to do so because it lacks the resources needed to guide or direct the behavior of other states in the international arena. Moreover, deficient international leadership is a feasibility constraint on the achievement of just climate action. Notwithstanding the strength of the moral arguments, if there are no actors who are able to take on the role of leader, it will be impossible to fulfill the role climate ethicists argue thos actors should take. This suggests there are good reasons to take a closer look at the nature of this particular feasibility constraint. If we have a better understanding of what kind of feasibility constraint that deficient international leadership presents, then we will better understand how to minimize its effects.

In this chapter, we will take a closer look at climate change leadership in the international arena. This will be done through a review of recent research on climate change leadership, including the identification of important gaps in the research. One such neglected area concerns the demand side of the leadership equation (cf. Karlsson et al. 2011). Leadership implies a relationship between leaders and followers, but studies of leadership often focus exclusively on the prospective leaders and their strategies, motives, or arguments. But in order to determine what *effective* leadership entails, we need to understand *who* potential followers recognize as a leader, as well as *why* an actor becomes recognized as a leader. That is, we need to understand what factors determine leadership recognition by potential followers. By thoroughly studying both the supply side and the demand side of climate change leadership we might reach a better understanding of how the leadership deficiencies emerge, and hence also better understand the nature of this feasibility constraint.

Our discussion will be illustrated by a case study in which we investigate both the supply side and the demand side of international leadership on climate change-induced human mobility. The purpose of examining this case is twofold. First, we think the case is interesting in itself. As far as we know, there are no previous studies looking at leadership in the context of negotiations on climate change-induced human mobility. Second, the case also provides both an illustration and a template of how one could establish how leadership has failed in a particular context. This also serves as a feasibility assessment. As Lawford-Smith (2013) points out, what we want to know when we ask about feasibility is whether there is an agent somewhere in the world who can bring a desirable outcome about. If there is, the outcome is feasible. If there is not, the outcome is infeasible. In order to identify agents and their possible actions, we need to do empirical studies. And when an outcome seems infeasible, we also need empirical studies to determine how and why it is so, so that we might, hopefully, do something about it.

I. FEASIBILITY AND INTERNATIONAL LEADERSHIP

In her thorough analysis of political feasibility, Lawford-Smith (2013) understands an outcome as feasible if there exists an actor (an individual or collective agent) with an action in their option set that has a positive probability of bringing the outcome about. The outcome of interest in this chapter is, roughly, a just international policy response aimed at developing solutions to address the challenges associated with climate change-induced mobility. We will, however, bracket the issue of what would, more precisely, constitute a *just* international policy response.[1] This is because we want the method for studying leadership deficits as feasibility constraints, which we will present in the following section, to be neutral between different understandings of what would constitute just policy responses. We do assume, however, that *some* form of *international* policy response is necessary.

We know from previous studies of international negotiations aimed at producing these kinds of policy responses that leadership is a critical determinant of success (Young 1991; Underdal 1994). Thus, feasibility assessments at the international level must examine leadership and see whether there is an actor with a leadership action in their option set that has a positive probability of bringing about an international policy response. The fact that there exists a leader is, of course, no guarantee that there will be an outcome, let alone a just one. But effective leadership seems to be very important to producing outcomes in international settings. So it is paramount that we have a thorough understanding of how deficient leadership can pose a feasibility constraint on just action.

What does it mean to say that there is, or is not, an actor who can bring a certain outcome about? An important aspect of feasibility concerns accessibility (Cohen 2009; Gilabert and Smith 2012). For an outcome to be feasible, there must be a way for an actor to bring it about; there must be what Buchanan (2004, 61) calls a "practicable route" from where we are now to the outcome we wish to reach. Gilabert and Lawford-Smith (2012) talk about accessibility in terms of a "trajectory." They point out that questions about whether an outcome is accessible and how accessible it is are questions about whether there is a trajectory, or a path of action, that an actor can take to arrive at the desired outcome. And, once again, when the outcome in question is an international policy response, the practicable route, or trajectory, entails an effective exercise of leadership on the part of some actor.

What kinds of constraints determine whether a trajectory, or route, is available? Gilabert and Lawford-Smith (2012) distinguish between hard and soft constraints. Hard constraints are logical or nomological. They serve as constraints in a binary sense: they either rule out a path of action or not. But when we discuss policy outcomes, the soft constraints are more relevant. Clear examples of soft constraints are economic, institutional, and cultural constraints. What is important about soft constraints is that they are neither permanent nor absolute. This means that they are malleable. Even if they pose difficult constraints on some course of action right now, they might not do so in the future. As Gilabert and Lawford-Smith point out, "[t]echnological innovations, political revolutions and radical cultural transformations are historically familiar phenomena" (Gilabert and Lawford-Smith 2012, 814). The fact that soft constraints are malleable is important. It means that, while we acknowledge that their impact is real, we can "imagine ways progressively to overcome those that block the pursuit of morally desirable goals" (Gilabert and Lawford-Smith 2012, 815). This highlights the significance of designing empirical studies aimed at understanding the precise nature of soft feasibility constraints. Only with a thorough understanding of a feasibility constraint can we start to imagine ways in which it could be overcome.

II. A TEMPLATE FOR STUDYING THE SUPPLY AND DEMAND OF LEADERSHIP

Drawing on philosophical studies of feasibility, we have constructed a template that provides a method for thoroughly studying a particular leadership deficit as a *feasibility constraint* on the achievement of just action. As we said earlier, we bracket the issue of what would, more precisely, constitute *just* action because we want our method to be neutral between different understandings of what justice would require.

The template consists of a four-step research process that examines both the supply and demand sides of leadership. By going through this process, it will be possible to diagnose where exactly the leadership deficit arises. We will also be able to identify what kind of constraint the deficit poses upon achieving just outcomes, and hopefully this will help us design strategies to overcome the constraints. Our template consists of four research questions:

I. Is there any actor displaying leadership?
II. If there is an actor displaying leadership, what kind of leadership?
III. Is the identified actor recognized by relevant others as a leader?
IV. What kind of leadership is in demand by relevant others?

When the template is used the issue of what justice requires can enter at different stages and is determined by some standard independent from the template itself. We could, for instance, pursue a rather restricted investigation and choose to focus only on actors who are pursuing, or promising to pursue, what we determine to be just action. Or we could cast a wide net to investigate the general availability of leadership and let considerations of justice enter only when we have gone through all the stages of the template. We will take the latter approach when, later in this chapter, we show how the template can be used.

In the following, we will illustrate how our template can be applied with a two-part analysis. The first part is an analysis investigating the supply of leadership. We will use a text analysis as the primary research method by which we investigate if there is any actor showing leadership in the international effort to address climate-induced human mobility. The textual data and analytical framework for this part of the analysis will be presented below. The next part of the analysis looks at the demand side of leadership. Here we will study survey results from the 2018 United Nations Climate Change Conference (COP24) looking at leadership recognition and leadership style demand among conference participants.

III. CLIMATE CHANGE–INDUCED HUMAN MOBILITY

Current levels of greenhouse gas emissions are predicted to distinctively alter the living conditions on earth (IPCC 2018). Both scholars and policymakers assert that climate change could result in population displacement on a scale the world is presently ill-equipped to address (International Institute for Sustainable Development 2011). In response, several proposals for new, legally binding multilateral instruments have been presented in scholarly literature and in the international policy arena. None of these proposals has yet been championed by a politically powerful national government (Wyman 2013, 168). Also, there is still no consensus among international political actors on which strategies countries should adopt to address climate change–induced displacement and migration.

Several scholars have suggested that a multinational legal framework is crucial for effectively dealing with climate displacement and migration (see, e.g., Gibb and Ford 2012). Other scholars have argued that such a framework is difficult to construct because it is both practically and conceptually impossible to differentiate between people displaced due to climate change and those displaced due to other economic or environmental hardship (McAdam 2011). Despite these controversies surrounding *how* to respond to climate change–induced displacement and migration, there is a strong consensus on the need for an *international policy response* that develops solutions that can adequately address the various challenges that lay ahead.

The term "climate change-induced human mobility" has emerged in the literature as an umbrella term encompassing displacement, migration, and planned relocation (Cf. Sierra Club 2018, 1). This is also the way that we will use the term throughout this chapter.

IV. THE IMPORTANCE OF LEADERSHIP IN INTERNATIONAL NEGOTIATIONS

Leadership is a critical determinant of success or failure in international negotiations; in order to achieve success, a certain minimum of leadership is required (Young 1991; Underdal 1994). There are several case studies on leadership within international regimes in areas such as trade, environment, and nuclear proliferation. They demonstrate that leadership has been crucial for the development of these international regimes (Sandholtz and Zysman 1989; Winham 1989; Benedick 1991). The literature on leadership has also recognized that the greater the complexity of the issue, the greater the importance of leadership (Underdal 1994). Climate change-induced mobility certainly qualifies as a complex issue, since it involves most countries in the world (both countries of origin and countries of destination) and has implications for several levels of legislation at the international and domestic level.

Scholars have identified certain behaviors that suggest that an actor is exercising leadership. Parker, Karlsson, and Hjerpe (2015) draw upon existing leadership theories to formulate four possible modes of leadership in an international negotiation arena. These four styles indicate that an actor is performing some kind of leadership behavior and describe what type of leadership this actor is performing. The four different modes of leadership are:

Structural leadership, which refers to the "deployment of power-resources for the purpose of creating new incentives and changing the costs and benefits associated with different avenues for action in a particular issue area" (Parker, Karlsson, and Hjerpe 2015, 4–5). This is illustrated by the ability to motivate others either by creating incentives via payoffs or providing resources. Structural leadership can also be illustrated through coercive action.

Directional leadership "rests on taking unilateral action and is accomplished by the demonstration effects of leading by example. By making the first move, it is possible to demonstrate the feasibility, value and superiority of policy solutions" (Parker, Karlsson, and Hjerpe 2015, 5). This is illustrated by a willingness and commitment to act on one's words.

Idea-based leadership "is concerned with the problem of naming and framing and the promotion of specific policy solutions to collective problems"

(Parker, Karlsson, and Hjerpe 2015, 5). Proposing joint solutions to collective problems, as well as making efforts to set the agenda, are examples of this type of leadership. Parker, Karlsson, and Hjerpe (2015) identify two dimensions of idea-based leadership: one involves making new proposals and suggesting innovative solutions, and the other consists of making efforts to change perceptions regarding the problem at hand.

Instrumental leadership "relies on negotiating skill and seeks to put together deals that would otherwise elude participants" (Parker, Karlsson, and Hjerpe 2015, 5). Instrumental leadership is closely related to idea-based leadership as it seeks to develop proposals to find common goals but with more emphasis on the skill of an actor to bring others together. This could happen by acting as a mediator to bridge problems.

Different types of leadership might be at play in the same negotiating setting. Moreover, one actor may be able to provide more than one mode of leadership in the same process, even though these modes are derived from different sources and are exercised through different behavioral strategies. Underdal (1994) asserts that an optimal mix of these leadership modes needs to be in place in order to secure effective leadership and thus a successful outcome.

It is important to note that merely performing actions that correspond to a leadership style does not necessarily mean that the actor *is* a leader. This is because leadership is a "relationship between leader and followers" (Underdal, 1994, 181). This brings attention to the demand side of leadership. In order to be a leader, one must both show leadership and be recognized as a leader by others.

In the following, we will first give an account of how actors show leadership in the area of climate change-induced human mobility. Then, we examine whether the identified actors seem to be recognized as leaders by others. And finally, we examine what kind of leadership is in demand in the area of climate change-induced human mobility.

V. FINDING LEADERS: THE SUPPLY SIDE OF LEADERSHIP IN INTERNATIONAL NEGOTIATIONS ON CLIMATE-INDUCED HUMAN MOBILITY

The first question in our template is, "Is there any actor displaying leadership?" This helps to identify which actors to study. To select actors, we assume that the probability of discovering leadership behavior increases if the actor has previously been identified as important in the literature. Thus, in order to identify which actors to select for closer assessment, a brief overview of the existing literature on the topic of climate-induced human mobility is required. Many of the reports on climate change-induced human mobility make little or no mention of specific actors. Further, most scholars who have studied actors in this area focus on nonstate actors such as international organizations, the research community, or nongovernmental organizations (see, e.g., Warner 2012; Hall 2013; Rosenow-Williams and Gemenne 2016). We will focus instead on the role of state actors, which have not been previously studied in this context.

We begin our brief overview of the literature on the topic of climate-induced human mobility by examining a report (IOM 2018) from the workshop "Climate change and human mobility: toward dignified, coordinated and sustainable responses" that took place in Morocco in May 2017. The report points out that the International Organization for Migration (IOM) and the United Nations High Commissioner for Refugees (UNHCR) are organizations that have significantly contributed to the discussion on climate change-induced human mobility. It also mentions several countries that have shown interest in the fate of persons displaced due to climate change, such as Switzerland, Norway, Bangladesh, and Morocco. The Alliance of Small Island States (AOSIS) is also acknowledged for doing extensive advocacy work in the effort to include the issue on the international political agenda.

Norway and Switzerland chaired the Nansen Initiative from 2012 to 2015, and have also been identified by scholars as actors playing crucial roles in the discussion on climate change and human mobility (see, e.g., McAdam 2014, 2016; Kälin 2012). As early as 2007, Norway encouraged UNHCR to "turn its attention to the issue of environmental degradation as a consequence of climate change" (Hall 2013; McAdam 2014). Norway and Switzerland's dedication to the issue of human mobility in the context of climate change was seen in 2011, when UNHCR proposed that it become the lead agency for coordinating protection responses in natural disasters. Only five countries indicated that they were willing to support UNHCR on this proposal, while all other states rejected this idea because they did not want to be governed by institutional actors (McAdam 2016). Furthermore, Norway and Switzerland did subsequently make a joint pledge "to cooperate with interested states, UNHCR and other actors with the aim of obtaining a better understanding of such cross-border movements." This pledge was also endorsed by Costa Rica, Germany, and Mexico (Kälin 2012).

Based on this overview, we will analyze the leadership behavior of three actors: AOSIS, Bangladesh, and Norway. This selection is based on an examination of important actors in the environmental migration nexus, and on how frequently these actors were mentioned in other literature relating to different aspects of climate-induced human mobility.

The second step of our template is to answer the following question: "If there is an actor displaying leadership, what kind of leadership?" To answer this question, we perform a text analysis that will enable us to tell whether the identified actors can be said to show leadership in the area of climate-induced human mobility, and, if this is the case, to provide an account of what kind of leadership the actors are excising. The text documents used in this study are publicly available documents containing statements made by the actors or documents issued by the actors themselves. To limit the number of documents subject to analysis, and to keep the analysis as up-to-date as possible, we will only consider documents released between the years 2013 and 2018.

This method of investigation comes with a risk, of course, since there can be gaps between written text and actual practice. An actor can state that they have accomplished P, but this does not necessarily mean that they really accomplished P. Even so, this method still tells us something about the leadership being offered by the "supplier." This justifies, we believe, using documents in which the actors are free to describe their own (potential) behavior and are therefore unrestricted in how they define their supply of leadership.

The framework used for analyzing our actors is based on the four modes of leadership outlined above, with each mode measured via the answers to three questions. Each question is addressed by identifying formulations or phrases in the selected texts that help us answer the questions with a "yes," a "no," or a "partly":

Structural leadership
Q1) Are there any remarks that suggest an attempt to exercise coercive action to achieve a certain goal?
Q2) Is there any evidence illustrating that the actor has the ability to provide resources?
Q3) Are there any examples of attempts to generate incentives via payoffs to achieve a certain goal?

Directional leadership
Q1) Are there any examples of attempts to lead by example?
Q2) Are there any examples where the actor is a driving force for ambitious results?
Q3) Is there any evidence that the actor has implemented a policy on climate change-induced human mobility?

Idea-based leadership
Q1) Is there any evidence of discovering and proposing joint solutions to collective problems?
Q2) Are there any examples of raising awareness on questions related to climate change-induced migration?
Q3) Is there any evidence of suggesting new innovative solutions and ideas?

Instrumental leadership

Q1) Are there any examples of a capacity to gather other actors in order to cooperate on issues?
Q2) Are there any examples of acting as a mediator to bridge problems?
Q3) Are there any examples of negotiation skills?

Each of the actors we have identified have been assessed separately using this analytical framework. When it comes to AOSIS, the analysis includes not only statements and actions from AOSIS but also statements issued by AOSIS member states in multilateral documents. Because of space restrictions, we will not present our full analysis here. Instead, we will present brief summaries of our findings for each actor, and a summary in table 7.1. The full analysis, and a full list of the documents we have analyzed, can be found at https://jodalen.wordpress.com/supplementary-materials.

The AOSIS

Between 2013 and 2018, AOSIS actively demonstrated directional, idea-based, and instrumental leadership in the international effort to address climate change-induced human mobility. AOSIS also seems to act as a driving force for ambitious results and made a significant contribution to the effort in raising awareness of climate change-induced migration. Even so, there was no clear evidence of AOSIS providing structural leadership regarding climate change-induced human mobility.

Bangladesh

In the documents we analyzed, Bangladesh showed clear directional and idea-based leadership in the international effort to address climate change-induced human mobility. When it comes to directional leadership, Bangladesh was an actor able to lead by example thanks to the various policies the country has developed in order to address internal climate change-induced displacement. However, there was no evidence of Bangladesh being able to perform structural leadership, and there was little evidence that Bangladesh has demonstrated instrumental leadership, other than gathering actors to collaborate on issues.

Norway

Norway primarily adopted the idea-based leadership style by making efforts to raise awareness not only about climate change-induced migration but also through proposing joint solutions to collective problems. Although Norway did not show any proof of implementing climate change-induced migration policy, they did show some directional leadership, mainly through acting as the driving force for ambitious results, and to some extent, through leading by example. There was also evidence of Norway deploying instrumental leadership, though this was mainly done through gathering actors within the Nansen Initiative and taking a lead in the UN towards improving the protection for internally displaced people. Norway did show structural leadership to some extent, mainly by providing financial resources to others.

Table 7.1   Leadership Supply

Leadership Mode	AOSIS	Bangladesh	Norway
Structural			
Q1: Coercive action	No	No	No
Q2: Provide resources	No	No	Yes
Q3: Generate incentives	No	No	No
Directional			
Q1: Lead by example	Partly	Yes	Yes
Q2: Ambitious results	Yes	Yes	Yes
Q3: Implement policy	Yes	Yes	No
Idea-based			
Q1: Joint solutions	Yes	Yes	Yes
Q2: Raising awareness	Yes	Yes	Yes
Q3: Innovative solutions	Yes	Partly	Yes
Instrumental			
Q1: Gather actors	Yes	Yes	Yes
Q2: Bridge problems	Yes	No	No
Q3: Negotiation skills	Yes	No	Yes

FINDING FOLLOWERS: THE DEMAND SIDE OF LEADERSHIP IN INTERNATIONAL NEGOTIATIONS ON CLIMATE-INDUCED HUMAN MOBILITY

We utilized a survey to study the demand side of leadership in this context. Our approach is strongly influenced by a method developed by Karlsson et al. (2011; 2012) in their studies of perceptions of leadership among climate change negotiation participants (see also Parker et al. 2012; Parker, Karlsson, and Hjerpe 2015; Parker and Karlsson 2018). To have an opinion on whether there are leaders in this area, survey respondents must have some basic knowledge about the problem at hand, as well as some insight into the behaviors of different actors present in the climate change-induced human mobility effort. Thus, we distributed our survey at the COP24 UN climate change conference in Katowice, Poland. We sought COP24 participants who were interested in issues related to climate change-induced human mobility by attending relevant side events and negotiations. Our selection strategy resulted in a total of 100 completed surveys.

Our survey utilized two questions. First, the respondents were simply asked to indicate which actor(s) (countries, party groupings, and/or organizations) they thought had a leading role in the negotiations on climate change-induced human mobility. The question was open-ended, allowing respondents to put down none, or as many actors as they believed to have a leading role. Second, the respondents were asked to indicate how they thought leadership in the area of climate change-induced human mobility is best demonstrated by specifying their level of agreement with six different statements, representing the four different leadership modes outlined above. As directional and idea-based leadership are a bit more complex than structural and instrumental leadership, we asked two questions about those, instead of one.

Our survey results suggest that there is no obvious actor with a strong leading role in the negotiations in this area. The answers are quite varied, and many actors are only mentioned once. Table 7.2 shows all actors who were mentioned by 2 percent or more of our survey respondents.

Table 7.2 Leadership Recognition. Responses to the question "Which countries, party groupings and/or organizations have a leading role in the negotiations on climate change-induced human mobility?"

Answer	Total	%	Answer	Total	%
Country Groupings:			*No Answer:*	33	15.5
AOSIS	17	8.0	N/A		
EU	16	7.5	*Intergovernmental Organizations:*		
SIDS	14	6.6	UNHCR	8	3.8
Countries:			UN	6	2.8
USA	10	4.7	UNFCCC	6	2.8
Bangladesh	7	3.3	IOM	5	2.3
China	6	2.8			
Fiji	5	2.3			

Note: Only actors representing ≥2 percent of the answers are included in this table. For a full account of all actors mentioned, see https://jodalen.wordpress.com/supplementary-materials.

Our results demonstrate two main trends. The first observation is that the largest share of the answers (15.5 percent) do not mention any actor at all because the survey question was left blank or respondents said they had no answer to the question. The second trend observed is that the small island states, AOSIS, and Small Island Developing States (SIDS) are mentioned in 14.6 percent of the answers. This share does not include the mentions of individual members of AOSIS and SIDS, such as Fiji (2.3 percent), the Maldives (0.9 percent), and Tuvalu (0.5 percent). This suggests that even if there is no obvious strong leader in the negotiations on climate change-induced human mobility, there is one conjoint actor—AOSIS/SIDS—which at least when compared to other candidates stands out as being recognized for taking a leading role by a notable share of the respondents.

Actors considered to be leaders in the general area of climate change negotiation (the European Union, United States, and China) were also represented in the answers, but not to the same extent as AOSIS/SIDS. Bangladesh was also mentioned as an actor having a leading role by a few (seven) respondents. Norway was mentioned by only one respondent. It should also be noted that intergovernmental organizations, civil society organizations, and other groups were mentioned by several respondents.

The survey results also tell us which leadership style the respondents considered to be the best in the area of climate change-induced human mobility. The results are presented in table 7.3. The mean indicates the respondents' average level of disagreement or agreement with the statements on a scale of 1–7, where one (1) means "disagree strongly" and seven (7) means "agree strongly."

Table 7.3 Leadership Style Demand.

Leadership Style	Mean	SD
Using resources to motivate others to take action (structural)	5.37	1.41
Providing a role model, implementing best practices that others can copy (directional)	5.67	1.35
Using resources to take significant domestic climate migration action (directional)	5.68	1.42
Providing visions (idea-based)	5.00	1.41
Developing new solutions (idea-based)	5.48	1.35
Establishing joint actions and initiatives (instrumental)	6.03	1.09

Answers to the question "Leadership on climate change-induced human mobility is best demonstrated by . . . " Level of agreement on a scale of 1–7.

The results show us that, in general, respondents believe that leadership on climate change-induced mobility can be demonstrated using all leadership styles because on average most respondents did agree or partly agree with all the statements. However, two statements stand out. First, respondents did on average agree more with the statement "establishing joint actions and initiatives," which represents instrumental leadership (with a mean of 6.03). Second, they did on average agree less with the statement "providing visions," which represents the idea-based leadership style (with a mean of 5.0).

VI. CONCLUDING DISCUSSION

So what do our results tell us about how leadership is demonstrated and recognized in the international effort to address climate change-induced human mobility? We have seen that between 2013 and 2018 all three actors analyzed on the supply side—AOSIS, Bangladesh, and Norway—demonstrated behaviors corresponding with most of the leadership modes. This means that all the actors we studied demonstrated behaviors that would qualify as exercising some sort of leadership. We also see, however, that the three actors differed with regard to the leadership style they deployed. Norway was the only actor who met even one of the indicators for structural leadership. AOSIS was the only actor found to act as a mediator to bridge problems, which is one of the criteria for instrumental leadership. The reason the actors differ in their behavior could plausibly be explained by their different incentives to act in certain ways, or by their different capacities. For example, it would be difficult for AOSIS and Bangladesh to demonstrate structural leadership as their resources to do so are limited.

Directional and idea-based leadership seem to be the two leadership styles most prominent among the analyzed actors. When it comes to idea-based leadership, all three indicators were met by both AOSIS and Norway, while Bangladesh met all but one. Regarding directional leadership, AOSIS and Bangladesh met all indicators, while Norway met all but one. It seems that directional and idea-based leadership are the modes that best characterize the supply side of leadership in the area of climate-induced human mobility. Structural leadership, however, is significantly missing in this area. We can conclude that the stated actions and behaviors of the studied actors show that there is indeed a supply of leadership in the international effort to address climate change-induced human mobility. But we also see that there are some important aspects of leadership, specifically structural leadership, that are not supplied to any significant extent.

Our next question is whether this supplied leadership is something that is noticed and recognized by others. Generally, our survey results indicate that the answer is "no." There seems to be a lack of actors recognized as leaders in the international effort to address climate change-induced human mobility, as many of our survey respondents did not mention any actor at all when prompted. Even so, one actor—AOSIS/SIDS—was mentioned more frequently than any other. AOSIS/SIDS is, of course, a coalition of countries very vulnerable to the impacts of climate change, which is probably the reason that they are ambitious in the negotiations on climate change-induced human mobility and why they are recognized by others. Our survey results also indicate that Bangladesh's supply of leadership is recognized by others, although not to the same extent as the supply by AOSIS/SIDS. The same explanation for why AOSIS is recognized is quite probably true for Bangladesh: it is a country very vulnerable to climate change. Norway, on the other hand, was only mentioned once in the survey. This indicates that Norway is not recognized as a leader in the area of climate-induced human mobility. We need to be careful, of course, not to overstate our conclusions. More studies need to be conducted before we can make any final judgments on these matters.

Our results also showed us which leadership style the respondents thought best demonstrated leadership in the area of climate change-induced human mobility. Although we could not identify any radical differences in the demand for different leadership styles—all styles were on average considered to be good for demonstrating leadership—we saw a somewhat higher demand for instrumental leadership. This result is interesting in relation to the results from the text analysis, where all of the actors analyzed did show a behavior *of gathering actors to collaborate on issues*. It seems then, that all actors were able to supply, to some extent, the type of leadership most demanded from others. Yet, only AOSIS met *all* criteria for showing instrumental leadership. If *gathering actors to collaborate on issues* can be said to represent the mode of instrumental leadership, the results from our text analysis would suggest that AOSIS is the actor that best meets the demand for leadership.

Even though all leadership styles were demanded to a fairly equal extent, it is problematic that none of our identified leadership suppliers seemed to supply substantial structural leadership. Parker, Karlsson, and Hjerpe (2015, 12) argue that providing good ideas, setting a good example that others can follow, or solving problems in negotiations, are important leadership tools, but in isolation they are unlikely to be sufficient to produce positive outcomes or widespread leadership recognition. Idea-based, directional, and instrumental leadership usually "need to be accompanied by the bargaining leverage that stems from structural power" (Karlsson et al. 2012, 48).[2] Similarly, Underdal (1994) states that an optimal mix of leadership modes needs to be in place in order to secure effective leadership, and Young (1991, 303) argues that if the efforts of a leader are to be successful, it will usually take the "contributions of different forms of leadership."

It is also problematic that Norway, the only actor demonstrating all modes of leadership, was not recognized as a leader at all in the area of climate-induced human mobility. Moreover, the actors recognized as leaders by at least a fair share of the respondents—AOSIS/SIDS and, to a lesser extent, Bangladesh—are relatively weak when it comes to central aspects of leadership style supply.

We are now able to say something more definite about the leadership deficit in this area. There *are* actors willing to take on the role as leader, but only one of them is recognized by others as a leader to any significant extent. Further, the actor recognized as leader does not supply any structural leadership. This is not necessarily a problem. If an actor is recognized as a leader despite the fact that it does not display a particular form of leadership, then this could mean that the particular form of leadership is not important. This would be a reasonable interpretation if the actor in question *is* recognized as a leader by a fairly large share of the potential followers, perhaps in combination with a low demand for the unsupplied mode of leadership. But as we already noted above, in the case we are studying here, structural leadership is in demand to about the same extent as the other forms of leadership. And while AOSIS/SIDS was the actor recognized for having a leading role by more respondents than any other candidate, it was still recognized by *only* 14.6 percent of the respondents. Most respondents did not mention any actor at all.

What can our analysis of leadership deficits as feasibility constraints say about the feasibility of international action aimed at addressing climate change-induced mobility? Let us remind ourselves of Lawford-Smith's (2013) understanding of feasibility: an outcome is feasible, she argues, if there exists an actor with an action in their option set that has a positive probability of bringing the outcome about. As we have seen, there are actors ready to supply leadership in negotiations on climate change-induced mobility. This means that there are actors who, at least potentially, could have actions within their option sets that promise to bring just outcomes about in this area. But as we have also seen, there is a combination of a lack of recognition of these actors as leaders on the demand side, together with a weak supply of structural leadership. Consequently, it is very doubtful that any of these actors would have a positive probability of bringing any just outcomes about. Given that effective political leadership is a crucial determinant for the realization of international agreements, our analysis suggests that just action to address challenges related to climate change-induced human mobility is currently infeasible.

That the only actor recognized as a leader to any substantial extent—AOSIS/SIDS—fails to supply structural leadership is not surprising. Structural leadership is about deploying power resources, or even engaging in coercive action, in order to change the costs and benefits associated with different courses of action. These kinds of power resources are largely unavailable to small and/or developing states such as the members of AOSIS (cf. de Águeda Corneloup and Mol 2014). As Karlsson et al. (2011) point out, it is often difficult for any single actor to successfully play all different required leadership roles. Ideally a number of actors would join forces to form a "leadership complex" that could efficiently confront the challenges at hand (ibid., 105).

Is there anything that can be done in order to reduce this feasibility constraint on developing solutions to adequately address the challenges of climate-induced human mobility? Others are better situated than us to suggest policy. But we wish to recall something we said earlier in this chapter. Leadership is a soft feasibility constraint and is malleable; the fact that there are no actors *right now* who seem to be able to fully exercise the kind of leadership in demand does not preclude that there might be such actors in the future. Gilabert (2009) stresses the importance of taking transitional standpoints, from which actors facing soft feasibility constraints envisage trajectories of reforms that promise to dissolve those constraints. Finding these transitional standpoints will require institutional experimentation. With a better understanding of the precise nature of a leadership deficit in place, we are also better prepared to formulate such potential transitional standpoints for the identified actors. What could Norway do, for instance, in order to become recognized as a leader in the area of climate-induced human mobility? Or what could AOSIS do in order to strengthen its structural leadership? These are the kinds of more precise questions we can ask now that we have a detailed diagnosis of the existing leadership deficit.

NOTES

1. We have made some suggestions on what would constitute just policy responses to address challenges associated with climate change-induced migration in Ödalen (2014) and Heyward and Ödalen (2016).
2. As Karlsson et al. (2012) also point out, it is equally problematic if an actor relies solely on structural leadership because it is difficult for a single actor to muster the power resources needed to make this a viable strategy.

REFERENCES

de Águeda Corneloup, Inés, and Arthur P. J. Mol. 2014. "Small Island Developing States and International Climate Change Negotiations: The Power of Moral 'Leadership.'" *International Environmental Agreements* 14(3): 281–97.

Benedick, Richard E. 1991. "Protecting the Ozone Layer: New Directions in Diplomacy." In *Preserving the Global Environment: The Challenge of Shared Leadership*, edited by Jessica Tuchman Mathews, 112–53. New York: WW Norton.

Buchanan, Allen. 2004. *Justice, Legitimacy, and Self-Determination*. Oxford: Oxford University Press.

Cohen, Gerry. A. 2009. *Why Not Socialism?* Princeton: Princeton University Press.

Gibb, Christine, and James Ford. 2012. "Should the United Nations Framework Convention on Climate Change Recognize Climate Migrants?" *Environmental Research Letters* 7(4): 045601.

Gilabert, Pablo. 2009. "The Feasibility of Basic Socioeconomic Human Rights: A Conceptual Exploration." *Philosophical Quarterly* 59(237): 659–81.

Gilabert, Pablo, and Holly Lawford-Smith. 2012. "Political Feasibility: A Conceptual Exploration." *Political Studies* 60(4): 809–25.

Hall, Nina. 2013. "Moving Beyond its Mandate? UNHCR and Climate Change Displacement." *Journal of International Organization Studies* 4(1): 91–108.

Heyward, Clare, and Jörgen Ödalen. 2016. "A Free Movement Passport for the Territorially Dispossessed." In *Climate Justice in a Non-ideal World*, edited by Dominic Roser and Clare Heyward, 208–26. Oxford: Oxford University Press.

International Institute for Sustainable Development. 2011. *Nansen Conference on Climate Change and Displacement Bulletin* 189(1). http://enb.iisd.org/crs/climate/nansen/html/ymbvol189num1e.html.

IPCC. 2018. "Summary for Policymakers." In *Global Warming of 1.5°C. An IPCC Special Report on the impacts of global warming of 1.5°C above pre-industrial levels and related global greenhouse gas emission pathways, in the context of strengthening the global response to the threat of climate change, sustainable development, and efforts to eradicate poverty*, edited by Valérie Masson-Delmotte, Panmao Zhai, Hans-Otto Pörtner, Debra C. Roberts, James Skea, Priyadarshi R. Shukla, Anna Pirani, Wilfran Moufouma-Okia, Clotilde Péan, Roz Pidcock, Sarah Connors, J. B. Robin Matthews, Yang Chen, Xiao Zhou, Melissa I. Gomis, Elisabeth Lonnoy, Tom Maycock, Melinda Tignor, and Tim Waterfield. Geneva: World Meteorological Organization.

IOM. 2018. *Analytical Report of the GFMD Workshop on "Climate Change and Human Mobility. Towards Dignified, Coordinated and Sustainable Responses."* https://environmentalmigration.iom.int/analytical-report-gfmd-workshop.

Kälin, Walter. 2012. "From the Nansen Principles to the Nansen Initiative." *Forced Migration Review* 41: 48–9. https://www.fmreview.org/sites/fmr/files/FMRdownloads/en/preventing.pdf

Karlsson, Christer, Charles Parker, Mattias Hjerpe, and Björn-Ola Linnér. 2011. "Looking for Leaders: Perceptions of Climate Change Leadership Among Climate Change Negotiation Participants." *Global Environmental Politics* 11(1): 89–107.

Karlsson, Christer, Mattias Hjerpe, Charles Parker, and Björn-Ola Linnér. 2012. "The Legitimacy of Leadership in International Climate Negotiations." *Ambio* 41, suppl. 1: 46–55.

Lawford-Smith, Holly. 2013. "Understanding Political Feasibility." *Journal of Political Philosophy* 21(3): 243–59.

McAdam, Jane. 2011. "Swimming Against the Tide: Why a Climate Change Displacement Treaty is Not the Answer." *International Journal of Refugee Law* 23(1): 2–27.

McAdam, Jane. 2014. "Creating New Norms on Climate Change, Natural Disasters and Displacement. International Developments 2010–2013." *Refuge: Canada's Journal on Refugees* 29(2): 11–26.

McAdam, Jane. 2016. "From the Nansen Initiative to the Platform on Disaster Displacement: Shaping International Approaches to Climate Change, Disasters and Displacement." *University of New South Wales Law Journal* 39(4): 1518–46.

Ödalen, Jörgen. 2014. "Underwater Self-Determination: Sea-Level Rise and Deterritorialised Small Island States." *Ethics, Policy & Environment* 17(2): 225–37.

Parker, Charles F., Christer Karlsson, Mattias Hjerpe and Björn-Ola Linnér. 2012. "Fragmented Climate Change Leadership: Making Sense of the Ambiguous Outcome of COP-15." *Environmental Politics* 21(2): 268–86.

Parker, Charles F., Christer Karlsson, and Mattias Hjerpe. 2015. "Climate Change Leaders and Followers: Leadership Recognition and Selection in the UNFCCC Negotiations." *International Relations* 29(4): 434–54.

Parker, Charles F., and Christer Karlsson. 2018. "The UN Climate Change Negotiations and the Role of the United States: Assessing American Leadership from Copenhagen to Paris." *Environmental Politics* 27(3): 519–40.

Rosenow-Williams, Kerstin, and François Gemenne, eds. 2016. *Organizational Perspectives on Environmental Migration*. New York: Routledge.

Sandholtz, Wayne, and John Zysman. 1989. "1992: Recasting the European Bargain." *World Politics* 42(1): 95–128.

Shue, Henry. 2011. "Face Reality? After You!—A Call for Leadership on Climate Change." *Ethics & International Affairs* 25(1): 17–26.

Sierra Club. 2018. *Women on the Move in a Changing Climate: A Discussion Paper on Gender, Climate and Mobility*. https://www.sierraclub.org/sites/www.sierraclub.org/files/program/documents/Women-Climate-Report.pdf.

Underdal, Arild. 1994. "Leadership Theory. Rediscovering the Arts of Management." In *International Multilateral Negotiation. Approaches to the Management of Complexity*, edited by I. William Zartman, 178–97. San Francisco: Jossey-Bass Publishers.

Warner, Koko. 2012. "Human Migration and Displacement in the Context of Adaptation to Climate Change: The Cancun Adaptation Framework and Potential for Future Action." *Environment and Planning C: Government and Policy* 30(6): 1061–77.

Winham, Gilbert R. 1989. "The Prenegotiation Phase of the Uruguay Round." *International Journal* 44(2): 280–303.

Wyman, Katrina M. 2013. "Responses to Climate Migration." *Harvard Environmental Law Review* 37(167): 168–216.

Young, Oran R. 1991. "Political Leadership and Regime Formation: On the Development of Institutions in International Society." *International Organization* 45(3): 281–308.

Chapter 8

Feasibility and Justice in Decarbonizing Transitions

Ivo Wallimann-Helmer

Climate change is expected to lead to rising sea levels, higher frequency of natural hazards, extended phases of drought, and many other negative impacts. To minimize these threats, the Twenty-First Conference of the Parties (COP 21) in Paris agreed that the global mean temperature must be kept "well below 2°C above preindustrial levels" (UNFCCC 2015). For this target to be feasible, most climate models assume not only heavy cuts in emissions but also negative emissions via the large-scale capture and storage of carbon dioxide (K. Anderson and Peters 2016). This chapter investigates the considerations of ethics and justice that are relevant to the fair governance of the carbon capture and storage technologies required to produce negative emissions.[1] I argue that the governance of these negative emission technologies (NETs) demands the fair distribution of the additional burdens and risks associated with them and that their ethical evaluation requires the appropriate involvement of those most directly affected. The main argument justifying these claims relies on the infeasibility of achieving justice in the governance of NETs. However, the solutions recommended for dealing with this challenge also face some feasibility constraints. I demonstrate how challenges of feasibility shape the directions of moral thinking about this area of climate action.

This chapter is structured as follows. I first provide an explanation of why NETs are seen as inevitable in most climate models. This means that our ethical analysis is no longer concerned with the justifiability of these technologies but with their urgent governance, which raises critical questions of justice. Relying on the most common principles of climate justice, I then explain what I take to be the main considerations of justice for the fair governance of NETs. Building on these principles, I show why I believe that procedural fairness, that is, the democratic involvement of all those most directly affected by

NETs, to be the most essential in their governance. Procedural involvement is crucial because, due to feasibility constraints, the fair distribution of the burdens and risks associated with the implementation of NETs cannot be realized. However, procedural fairness is itself challenging to achieve because it means that, in many cases, historical borders of jurisdiction would have to be disregarded and that, at least in some socioeconomic milieus, education would have to be boosted strongly and very rapidly. I end with some more general speculations about what my argument means for other areas of climate policy in which action is taken, and injustices result.

I. THE INEVITABILITY OF GOVERNING NETS

In this section, I explain why, in light of the most recent developments in international climate politics, governance of NETs as one form of geoengineering is inevitable. I argue that we must investigate the various challenges associated with governing NETs because there is a high probability that many NET facilities will be built, most likely at the cost of socioeconomically disadvantaged peoples and members of society. In consequence, we need to investigate challenges of justice that arise in the implementation of NETs.

One of the most important outcomes of the 2015 negotiations of the Parties to the United Nations Framework Convention on Climate Change leading to the Paris Agreement was the agreement to keep the rise in global mean temperature "well below 2°C above pre-industrial levels and to pursue efforts to limit the temperature increase to 1.5°C above pre-industrial levels" (UNFCCC 2015). This is an important political breakthrough because this target sets a very low acceptable temperature rise before climate change is considered to become dangerous. As a consequence of this agreement, heavy emission cuts are necessary. They must be achieved by reducing the emission intensity of lifestyles, in energy production, and in industry. However, most climate models assume that reaching this goal without substantial overshoot in temperature increase is only possible by investing in geoengineering, or more precisely technological interventions that produce negative emissions (Geden 2016). Hence, with this goal a conditional feasibility constraint is established (Gilabert and Lawford-Smith 2012). Without the necessary technological development and implementation of NETs, the target of the Paris Agreement cannot be achieved.

In principle, technological intervention in the climate system takes two forms (Pasztor 2017). Either the albedo of the planet is enhanced to minimize the solar radiation that is warming the atmosphere or carbon is captured from the atmosphere. The first technological intervention aims to reduce global warming by reflecting sunlight, while the second intervention achieves the same end by reducing the accumulated concentration of carbon in the atmosphere. Climate models seldom assume that solar radiation management, the first variant of technological intervention, can be widely used to help reach the target of the Paris Agreement because the technology is far from mature and the risks of negative side effects are very high (Honegger et al. 2017). For these reasons, most climate models that predict the achievement of the Paris Agreement target assume the capture and storage of carbon at large scale.

Technologies for capturing carbon come in several variants and most often go together with permanently storing liquefied carbon underground (Minx et al. 2017). When they store more carbon than is emitted, they produce negative emissions and hence are NETs. NETs most prominently include afforestation, agriculture, and other land-use changes to capture carbon (AFOLU), bioenergy production with carbon capture and storage (BECCS), and direct air carbon capture and storage (DACCS). In an IPCC special report investigating the possibility of no more than 1.5°C warming above preindustrial levels, most climate models assume that either AFOLU or BECCS will produce negative emissions (IPCC 2018). Both these interventions demand vast amounts of agricultural land and water. Both afforestation and BECCs capture carbon in fast-growing biomass, most probably grown in monocultures. By contrast, DACCS capture carbon with filters directly from the air or where it is produced. Both BECCS and DACCS require facilities for energy production and liquefying carbon for it to be stored. Beyond these facilities, BECCS and DACCS include transportation of carbon and its storage in appropriate geological formations or in the deep sea. As I explain in the following, all these elements lead to ethical challenges along with feasibility constraints on addressing them.

To date, the ethical debate surrounding geoengineering has mainly focused on whether the implementation of geoengineering technologies can be morally justified (Pamplany, Gordijn, and Brereton 2020; Gardiner 2020). In contrast, and in consideration of the ambitious target of the Paris Agreement, the international political community accepted a very high probability of the large-scale implementation of geoengineering with NETs. International policy decisions entail that the main question is no longer whether NETs can be morally justified. Instead, the key question becomes how they can best be implemented and governed (Minx et al. 2018). However, there are at least two arguments from the debate on the ethical justifiability of geoengineering that remain important because they are pertinent to reaching the Paris target (Lenzi 2018). One is the argument that the promise of a technological solution to the climate challenge will reduce the preparedness to make major reductions in the production of emissions. The other is the risk that implementing NETs at the scale assumed by most climate models will not be technically feasible. Both these challenges put at risk the feasibility of reaching the target of the Paris Agreement.

The first challenge concerns what has been called the moral hazard argument (Hale 2012; Lin 2013). This argument refers to the phenomenon, documented in the insurance literature and in economics, that people tend to refrain from risk reduction measures if they know that insurance is available. In the case of geoengineering, technical intervention in the climate system is understood as analogue to insurance and failure to reduce emissions as taking higher risks. Despite the need for heavy cuts in the production of emissions for the Paris target to be feasible, the possibility of reducing carbon concentrations by technology may actually decrease efforts to change emitting behavior. The second challenge concerns the immaturity of the technological development of NETs (Lenzi, Lamb, William F., Hilaire, Jérôme 2018; Bui et al. 2018). Even though many NETs are advanced in their development, it is still highly questionable whether they can be implemented at the large scale, in the short time, and with the level of efficiency assumed in most climate models (Fuss et al. 2014). If such fast development and implementation of NETs is not possible, the Paris target seems not to be reachable. Consequently, governing NETs demands dealing with both these challenges as feasibility constraints. However, they are not conditional constraints but rather concern motivation and chances of success (Roser 2016; Wiens 2015).

While these first two challenges put into question the feasibility of achieving the Paris target, other challenges will arise independently of whether global warming is kept below 2°C or even 1.5°C above preindustrial levels. They concern other goals that are put into question by implementing NETs at the scale needed for the Paris target to be feasible. Take the NETs AFOLU and BECCS, the two technologies most widely assumed in climate models. Both these technologies demand vast amounts of agricultural land and water (Boysen et al. 2017). This demand can conflict with the protection of biodiversity and food security. Binding appropriate amounts of carbon requires fast-growing biomass that will most probably be grown in large monocultures and in tropical regions where the risk of biodiversity loss is very high (Dooley, Harrould Kolieb, and Talberg 2020). Moreover, the vast amounts of land and water necessary to realize NETs at the scale needed puts intense pressure on the competition for agricultural land. This will probably undermine food security in poor regions of the world because either land is no longer available or because the restriction on available areas will drive up the prices for basic staple foods (Williamson 2016).

Neither of these challenges concerns the feasibility of the Paris target as such but the feasibility of realizing the conflicting goals of negative emissions, biodiversity conservation, and food security. This is an ethical dilemma because promoting one of the three goals—either achieving the Paris target via negative emissions, conserving biodiversity, or promoting food security—probably means diminishing the chances of realizing either or both of the other two. Not realizing one of these goals would mean a moral failure, whereas realizing them all together seems unfeasible. However, even though this dilemma is a challenge, NETs like afforestation or BECCS to bind and store carbon underground or in the deep sea will be implemented anyway. The international community has decided the Paris target that requires the large-scale implementation of NETs. These implementations concern further ethical challenges, which can also lead to new feasibility constraints. The remainder of this paper investigates these constraints. They concern the just distribution of negative impacts from NETs. I argue that these constraints are normatively relevant because they should shape how we should realize the distribution of negative impacts from NETs (Southwood 2018).

II. UNFEASIBLE JUSTICE IN NEGATIVE EMISSIONS

As shown in the literature, NETs will have positive and negative impacts (Kortetmäki and Oksanen 2016; Fuss et al. 2018; Fogarty and McCally 2010). Beyond reducing the concentration of carbon, other positive impacts include job and other economic opportunities, and probably financial compensation for assuming the higher burdens and risks associated with the implementation and operation of such facilities. These higher burdens and risks establish the negative impacts of NETs. Burdens concern noise emissions due to the running of NETs, use of land and water by the facilities that can then no longer be used for local agriculture, and the simple increase in traffic because of additional individuals working at these facilities. Risks concern incidental harm due to potential leakages and other accidents that could have negative impacts on the health and quality of life of local people. Governing these positive and negative impacts in a fair way and clarifying what this means in institutional and distributive terms is crucial. In this section, I discuss the most pertinent principles for justifying a fair distribution of the negative impacts of NETs, and I also argue that feasibility constraints render such a distribution difficult, if not impossible.

Implementing NETs involves similar issues to those raised by waste policy and other areas in which environmental risks have to be distributed. Empirical environmental justice research often considers a situation where disadvantaged social groups bear unequal negative environmental impacts as an injustice (Walker 2010). However, for such a claim of injustice to be valid, it must be shown that an unequal distribution of burdens is in conflict with some relevant principle of distributive justice (Wallimann-Helmer 2019b). In the case of waste policy, principles of distributive justice define the conditions under which the geographical distribution of waste facilities in society can be deemed just. The same holds for the distribution of NETs, but these include both facilities to produce energy and liquify carbon like in case of BEECS and DACCS and the biomass plantations in the case of afforestation and BECCS. Furthermore, in the case of BECCS and DACCS liquefied carbon must be transported and stored in appropriate geological formations or in the deep sea. All these aspects of implementing NETs lead to additional burdens and risks for those living nearby and so must be governed in a fair way (Fuss et al. 2018).

At least three principles of climate justice may become relevant in these contexts (Hayward 2012; Pozo et al. 2020): the equal-per-capita principle, the polluter pays principle, and the beneficiary pays principle. As I explain in more detail below, in contrast to their usual application as principles of climate justice, in the case of the distribution of the negative impacts of NETs, they must be specified slightly differently (Wallimann-Helmer 2019a). First, the equal-per-capita principle demands that all members of society should be equally exposed to the various burdens and risks from NETs. Second, the polluter pays principle (PPP) demands that exposure to the additional burdens and risks that result from NETs should be in proportion to the level of emissions produced by the relevant parties because it is the differing contribution to climate change that makes these facilities necessary in the first place. Third, for similar reasons, the beneficiary pays principle (BPP) demands that exposure to the negative impacts from NETs should correlate with the differing benefits extracted from the production of emissions by others. These specifications of the three principles differ from how they are used in the debate about climate justice (Gardiner et al. 2010). In this debate, none of these principles defines the fair distribution of risks. PPP and BPP serve the purpose of differentiating responsibilities for climate action, and the equal-per-capita principle defines entitlements to emitting.

In empirical environmental justice research, the most prominently invoked principle is the equal-per capita principle, more commonly known as the principle of equality. According to this principle, environmental risks should be distributed equally across society and all socioeconomic strata (Walker 2012). One common argument in defense of this principle is that equality is the default distribution when distributing goods and burdens (Gosepath 2015). This is because an equal distribution of goods and burdens is the only pattern of distribution not in need of justification. However, this presumption can be questioned in the case of NETs. Our different lifestyles do not all produce equal amounts of emissions and so do not contribute equally to the need for NETs. Therefore, it seems more plausible to demand that the negative impacts of NETs be distributed in accordance with the differing contributions to climate change. This is where the PPP comes in (Gardiner 2004).

According to this principle, a just distribution of negative impacts from NETs should be proportional to the contribution to climate change.

The PPP has some intuitive appeal. Most believe that those who produce a mess have a duty to tidy it up; in the case of climate change, this means that they should accept higher burdens of climate action (Shue 1999). This fits perfectly with a distribution of negative impacts from NETs that is proportional to contributions to climate change. However, emissions are produced not only by the use of goods but also by their production. Consequently, those consuming the goods do not necessarily directly produce them. If understood this way, the PPP can foster injustices. The affluent are usually those who consume more goods but do not work in the production of these goods and thus do not directly produce the emissions. According to the BPP, it is therefore not those who produce emissions that should face higher environmental risks; rather, it is those who benefit from the production of such goods (Page 2012). As such, the affluent, having the most emission-intensive lifestyles, are those who should be most exposed to the negative impacts of NETs.

According to all three principles of distributive justice mentioned thus far, an unequal burdening of the socioeconomically disadvantaged with the negative effects of implementing NETs seems unfair because the emissions they produce in the consumption goods, or the benefits they extract from emissions created in the production of goods, are most probably proportionally lower. For the global distribution of NETs, this means that developed countries and some emerging countries should shoulder the main share of NETs and their negative impacts (Fyson et al. 2020). At a national level, this would mean siting NETs in a way that corresponds to the emission intensity of domestic socioeconomic groups.

However, due to the technology, it is not possible simply to distribute NET facilities and their local negative impacts according to these principles of justice. Biomass must be grown where there is fertile land. Liquefied carbon must be stored in appropriate geological sites or in the deep sea. Hence, as plausible as PPP and BPP might seem at first sight, they can be challenged, because feasibility constraints render a just distribution of the negative impacts of NETs difficult if not impossible to achieve (Fogarty and McCally 2010; Steigleder 2017). In addition, with BECCS and DACCS, it is simply more efficient to gather biomass, produce energy, and liquefy carbon in a small number of facilities than to have many facilities spread all over the country. Logistical reasons justify siting afforestation areas and BECCS facilities in regions that are better suited for growing biomass. The same holds for the storage of liquefied carbon, for which security reasons are most pertinent. Furthermore, for technical reasons, centralized facilities are better suited to processing larger amounts of biomass or, in the case of DACCS, to filtering larger amounts of carbon from the air. Even though centralized facilities could be built away from populated areas, they will still only impact one part of society and not all emitters equally or proportionally to their benefits extracted from emissions.

Although efficiency, security, and technical concerns are crucial, they must be weighed against considerations of distributive justice, since they are insensitive to social and economic disadvantages. Communities living in geographical regions that allow more efficient growth of biomass or the production of energy and liquid carbon, or communities that are more secure or offer a more technically feasible storage of liquified carbon, cannot simply be said to be better able to accept the additional burdens of NETs. Such claims would amount to another principle of climate justice, the ability-to-pay principle (APP) (Caney 2010). This principle holds that those best able to provide solutions to climate threats are under a duty to shoulder heavier burdens. However, such considerations are highly questionable in the distribution of the negative impacts of NETs. As plausible as it might seem that those countries with the best know-how and resources to implement NETs should be responsible for shouldering more burdens, APP is less appropriate when it comes to the distribution of the negative impacts of NETs. Applying it cannot guarantee that there is no significant correlation between exposure to higher environmental risks and socioeconomic disadvantage.

III. THE RELEVANCE OF PROCEDURAL INVOLVEMENT

From the previous section, it appears that efficiency, security, and technical reasons make it highly probable that a just distribution of the negative impacts from NETs cannot be achieved, at least not in the foreseeable future. Two further considerations of justice can ameliorate this kind of injustice by legitimizing or increasing the social acceptability of such unjust siting: procedural justice and compensatory justice. Procedural justice, that is, the democratic participation in policy decisions about implementing NETs, could help legitimize the unequal burdening of certain parts of society. Conversely, compensating the local communities in which NET facilities are sited would allow the unjust distribution of their additional burdens and risks to be rectified. In the following, I argue that procedural justice must be guaranteed not only in policy decisions about implementing NETs but also in decisions about what kind of compensation is just for corresponding additional burdens and risks.

Empirical environmental justice research not only detects injustices in the distribution of environmental burdens and risks, but in many cases, it also finds that the communities exposed to higher environmental risks and burdens have not been properly involved in the political processes leading to the unequal distribution (Schlosberg 2007). Failing to involve those most directly affected by higher environmental burdens and risks in policy decisions is problematic for at least three reasons. First, failing to appropriately involve those potentially affected means discriminating against their interests unjustifiably. According to the all-affected principle, it is illegitimate to deny a voice in policy decisions to those who have morally relevant interests at stake (Goodin 2007).[2] Second, it is unacceptable to expose others to risks if they have not had the possibility to refuse (Hansson 2013). Third, there is an empirical argument why democratic involvement is important: sociopsychological research on nuclear waste repository siting in Switzerland has shown that transparency and the involvement of those affected can substantially increase the social acceptance of higher environmental burdens and risks (Krütli et al. 2015). In the case of the large-scale implementation of NETs, this last point may be critical: without increased social acceptance of facilities and storage sites for liquid carbon, implementation might become impossible.

These three arguments indicate three conditions of procedural justice applicable to policy decisions about implementing NETs (Callies 2018). First, according to the all-affected principle, all those most directly exposed to morally relevant negative impacts of NETs should be involved in decision-making about their governance. Second, it is important to secure appropriate scientific information and competence for all affected because only they can decide how to evaluate the risks they face (Wallimann-Helmer 2015). Third, to increase the social acceptance of siting decisions, it is key to make the rules for decision-making as transparent as possible. Taken together, these three conditions can be said to define the core of what it means to ensure procedural justice in governing not only NETs but also many other environmental risks. However, these conditions also face challenges.

As I argue elsewhere, the all-affected principle needs closer definition in two respects (Wallimann-Helmer 2019b). First, who is directly exposed to the negative impacts of NETs must be clarified. A way of doing so would be to define exposure by geographic proximity. The closer one lives to NETs, the more one is affected. Of course, such affectedness must be significant enough to become morally relevant. Showing this might become challenging because at least some studies argue that the negative impacts associated with NETs are not very high (Zhang and Huisingh 2017). Second, merely being exposed to some kind of change in burdens and risks is too vague to define morally relevant exposure. This is especially important considering the differences in the capacities of socioeconomic strata in society (Morello-Frosch et al. 2011). Those more socioeconomically disadvantaged will most probably suffer more from additional burdens and risks than those better off. Due to a lack of financial resources, they cannot afford to move away from places with higher environmental risks without great difficulty and exactly these places potentially also tend to attract poorer people (Pastor, Sadd, and Hipp 2001). At the same time, some studies indicate that socioeconomic disadvantage increases the risk of ill-health: social gradients in health show that economically vulnerable people face higher risks of ill-health than those better off (Brulle and Pellow 2006).

Providing appropriate information to all those who are affected in a morally relevant sense is crucial to ensuring procedural involvement under fair conditions. I discuss this issue in the next section. However, information is also important for enhancing the social acceptability of NETs. Social acceptability concerns not only the burdens and risks accompanying NETs but also for understanding their necessity for reaching the Paris target (Gough and Mander 2019). Realizing NETs at the scale assumed by most climate models requires implementing NETs in various geographical and societal circumstances in many different places; all of those potentially affected need to be convinced to accept the implementation of NETs in their neighborhood. For the same reason, it is also crucial to ensure transparent information about decisions concerning NETs (Krütli 2010). Without trust in the fairness of decision procedures, social perception of the legitimacy of these decisions will likely decline. Therefore, ensuring the appropriate democratic involvement of those most directly affected in a morally relevant sense by NETs becomes a crucial feasibility constraint. Moreover, without minimal consent to the implementation of NETs, the probability of their implementation in due time will surely diminish.

Increasing social acceptability also requires that the injustices resulting from the unequal distribution of the negative impacts of NETs are remedied. However, what compensation is appropriate as a remedy in these kinds of cases is not as straightforward as it might seem (O'Neill 2017). It depends not only on how the negative impacts can be determined empirically but also on the individual assessments of those facing these additional burdens and risks. Generally speaking, the goal of compensation should be that those affected subjectively feel as well off as before they were injured (Goodin 1989). The meaning of this claim is most straightforward in cases of monetizable harm. If someone is harmed and must be compensated, then it is appropriate that that person receives enough money to be able to pursue the same ends as before. In Goodin's (1989) terms, this is means-replacing compensation. In this case, money allows the means damaged or lost to be repaired or replaced. However, not all harm is monetizable, and it is not even clear in all cases of monetizable harm what amount of money makes good for the assets damaged or lost, and in some cases, the assets lost cannot be replaced at all. The ends usually realized by these assets must be modified. According to Goodin (1989), this demands another kind of compensation: ends-displacing compensation, such as assistance in changing livelihood due to loss of agricultural land.

This distinction between two kinds of compensation is especially important if NETs increase health risks or other risks of harm that are not readily monetizable. It is not straightforward what amount of money makes good for bad health or an increased risk thereof. It is difficult to say what amount of money appropriately compensates for the loss of livelihood due to, for instance, loss of agricultural land. In these kinds of cases, it might be better not to try to replace people's means but to compensate them by helping them change their objectives. That is, the best goal of compensation might be assisting in modifying ends so that those being compensated no longer depend on the means damaged or lost in order to feel subjectively as well off as before. Of course, such reasoning risks becoming overly paternalistic, prescribing to those facing losses how they should, for instance, change their livelihoods.

These considerations show why the involvement of legitimate claimants for compensation in implementing NETs is key. Such claimants should not only assess the tolerability of the additional burdens and risks they face but should also have a voice in determining what compensatory measures are most appropriate and, in order to avoid paternalism, in what way they should modify their objectives if need be. Admittedly, democratic involvement of all affected cannot outweigh an unjust distribution of the negative impacts of NETs, even if accompanied by appropriate compensation. At best, it can legitimize an unjust distribution of additional burdens and risks and, if fair and transparent decision procedures are applied, increase the social acceptability of NETs. However, since efficiency, security, and technical reasons render a just distribution of negative impacts from NETs very challenging, if not unfeasible, this will probably be the best approximation to a just distribution of negative impacts we can hope for.

IV. FEASIBILITY CHALLENGES FOR INCLUSIVELY LEGITIMIZING UNFAIRNESS

Following the argument thus far, unequal exposure of parts of society to the higher burdens and risks of NETs can become acceptable. This can result if those most directly affected have been appropriately involved in policy decisions related to the governance of the negative impacts of NETs and about compensation for additional burdens and risks. This demands that those affected can have an effective voice in the decisions concerning NETs. I first argue in this section that, depending on how effective and equal voice is defined, not only must socioeconomic inequalities be reduced to a greater or lesser extent but all the inequalities that are acceptable in the distribution of environmental risks also vary. However, this does not define who should be involved in policy decisions related to NETs. I argue, second, that the decision-making body must be flexible depending on how far the negative impacts of different NETs probably expand in time and space. As plausible as both these arguments may seem, they are actually quite radical and lead to two additional feasibility constraints in governing NETs.

In my view, the most promising approach to ensuring effective and equal voice in NET decision-making is to look at different understandings of democratic citizenship through the lens of justice (Schuppert and Wallimann-Helmer 2014). This lens shows that the acceptability of any deviation from a just distribution of negative impacts in governing NETs depends on our understandings of democracy and citizenship. We can think of democratic citizenship as either a formal or a substantial requirement of justice. Formal and substantial requirements demand different conditions for citizens to have an equal and effective voice in policy decisions. In doing so, they demand more or less extensive reduction of socioeconomic inequalities. This determines the space for unequal yet legitimate distributions of the negative impacts of NETs.

According to the formal requirement, citizenship is defined by the formal and legal rights every citizen enjoys (Downs 1957). These include the right to vote, the right to life, and all other rights usually considered basic rights. On this account, once a formal right to vote is granted, any policy decision is acceptable if it does not undermine basic rights. This means that policy decisions about NETs are acceptable as long as all citizens affected have a formal right to participate in the processes leading to those decisions and these decisions do not undermine their basic rights. However, a formal account faces two challenges. First, the socioeconomic inequalities between the parties involved in policy decisions need not be counterbalanced. Although these inequalities most probably increase the unfairness in the distribution of negative impacts from NETs, according to the formal requirement of citizenship, the right to vote is enough to ensure legitimate decisions. Second, under the condition that the resulting policy decisions do not infringe upon basic rights, they are deemed acceptable irrespective of how negative their consequences are for those exposed and most vulnerable to the negative impacts of NETs.

The substantial requirement of citizenship can remedy these unfavorable implications by demanding that policy decisions are only acceptable if citizens are substantially equal in their powers during policymaking and remain so thereafter. According to this account, unequal distributions of environmental burdens and risks can only be acceptable to the extent that they do not undermine substantial conditions of justice (van Parijs 2011). However, while the formal requirement leaves great leeway for inequalities, the substantial requirement is far too demanding. It claims that policy decisions are acceptable only if they are in line with comprehensive conditions of justice because according to this view democracy serves as a means to realize justice. As argued above, however, efficiency, security, and technical considerations render this infeasible. Therefore, for the implementation of NETs to be possible at all, the substantial requirement of citizenship should not be interpreted too restrictively and this is even more urgent considering the scale at which climate models assume NETs to be realized. Hence, what seems to be urgently needed is a middle-ground position incorporating both some formal and substantial requirements of citizenship in decision-making about governing NETs.

One version of this position accepts unjustifiable inequalities in the distribution of the negative impacts of NETs but demands that, to be fair, those on whom they fall have the formal and substantial conditions for free and equal citizenship (E. Anderson 1999; Scheffler 2015). If these conditions are not undermined, any inequalities in the distribution of any additional burdens and risks resulting from the decision procedures are acceptable. I take this conception of citizenship to be attractive because it secures at least some basic substantial equality between citizens while leaving enough leeway for acceptable policy decisions not fully aligned with principles of just outcome distribution. Following this conception of citizenship, it is appropriate that all those relevantly affected by the negative impacts of NETs are provided with some basic standard of substantial equality while being granted formal rights to participate in decisions about them. Any inequalities resulting from the distribution of the negative impacts of NETs should neither infringe upon these substantial conditions nor the formal rights to participate.

However, this conclusion does not determine the appropriate extension of the decision-making body, that is, who should be included in governance decisions about NETs. This is for two reasons. First, exposure to NETs and to their associated burdens and risks are not restricted to the boundaries of existing jurisdictions or a state. Even though many negative impacts of NETs might be rather local, leakages or other accidental harms in facilities close to borders will not be restricted to given jurisdictions. Second, burdens and risks from NETs do not exclusively concern those living at the time of policy decisions about their implementation and large-scale use but also future generations. For example, future generations are affected by potential leakages of deposits of liquid carbon or by the fact that NETs might not deliver the effects on the climate system that most climate models expect them to do. The first reason for inclusion deals with the relevance of geographical proximity to NETs and how to recognize morally relevant interests beyond the boundaries of given jurisdictions and states. The second reason concerns the inclusion of future generations in the policy decisions of those living today. Similar to the geographical challenge of inclusion, the interests of future generations tend to be ignored by those involved in policy decisions today. Consequently, both aspects of inclusion put into question how the decision-making body governing implementations of NETs should be defined.

The geographical challenge of inclusion may plausibly be answered by involving all those geographically exposed to the consequences of policy-making irrespective of boundaries (Valentini 2012). The challenge in this case will be to define who is affected in a morally relevant sense. As already discussed, a usual way to define morally relevant affectedness is by relying on morally relevant interests. These interests are those that concern our basic needs and rights as human beings and citizens. They also concern the conditions of fair procedural involvement. But the level of risk of infringement of these interests that are sufficient to require a say in decisions about NETs and their governance is not easy to determine. In principle, everyone is affected by NETs because if realized at a large enough scale, they influence the global climate. Furthermore, even though the effects may be small, everyone could claim to be affected by every policy decision made anywhere in the world. This so-called "butterfly effect" might justify introducing meta-deliberative processes that allow everyone who believes themselves to be affected by NETs in a relevant sense to demand inclusion in their governance (Fraser 2010).

Although not so difficult to imagine, including all those affected in a morally relevant sense in decisions about governing NETs is quite unusual for states as we know them today, because their boundaries of jurisdiction usually remain stable and do not vary depending on the policy issue. The lack of inclusion of future generations cannot be overcome by varying the decision-making body as currently constituted. What would be needed for their inclusion is some kind of proxy representation in the policymaking process and corresponding changes in government institutions (Wallimann-Helmer, Meyer, and Burger 2016). Hence, the challenge of the inclusion of future generations requires changing decision-making institutions by adding new institutions or representatives for those who do not yet exist to most existing government structures.

However, in most countries, institutional reform for both challenges of inclusion needs more time than is available to deal smoothly with all the pressing governance issues accompanying the large-scale implementation of NETs. Even though there might be occasional opportunities for immediate institutional changes through political deliberation in some countries, it is highly doubtful that the necessary changes can be realized in due time everywhere that NETs would have to be implemented (Niemeyer 2014). Furthermore, governing NETs not only demands building appropriate institutions to involve all affected but also their education and equal involvement in deliberative processes in a very short time, because models showing the Paris target to be possible assume the large-scale implementation of NETs rather soon. Hence, since substantial equal democratic citizenship means not only providing information but also education to all those affected in a morally relevant sense, this creates a further challenge of feasibility. The same holds for realizing institutional reforms in a short time. Overcoming rigid legislative boundaries as we are used to them today and establishing governance schemes that flexibly adapt to various ways of affectedness cannot be realized in due time. The same holds for institutions seeking to better include future generations.

As a consequence, this means that those deciding about NETs, their negative impacts, and probable compensatory measures, should be ready to recognize that the people potentially vulnerable to their decisions are not only those officially involved in the decision-making processes but also born or unborn others beyond the jurisdiction of the state involved. This makes it necessary to accept that there are always legitimate claimants for inclusion in decision-making, and procedures should be designed to ensure the potential inclusion of those not yet included, if appropriate (Fraser 2010). However, if such processes do not yet exist, there is a great danger that those excluded will not be able to make their voices heard. Accordingly, all those affected in a morally relevant sense are most probably not sufficiently included in the relevant decision-making body. This potentially undermines the legitimacy of these decisions and shows another feasibility challenge in governing NETs.

V. CONCLUSION

This chapter investigated the governance challenges arising from the large-scale implementation of NETs. I argued that such an investigation is crucial because most climate models assume that in order to keep the increase in global mean temperature to no more than "well below 2°C above pre-industrial levels," the implementation of NETs at such scale is necessary. Many NETs and their facilities will involve additional burdens and risks, mainly befalling individuals and communities living geographically near these kinds of facilities. This is why analyzing the justice implications of the distribution of their negative impacts is crucial to avoiding unfair inequalities such as those that have been observed for some time already in waste policy and with other kinds of environmental risks. However, as I have shown, there are efficiency, security, and technology constraints that render a fully just distribution of these additional burdens and risks difficult if not entirely infeasible. These reasons make it necessary to involve all those most directly affected by NETs in decisions about their governance. However, institutionalizing this claim also faces feasibility challenges. For such involvement to be fair and appropriate, radical institutional reforms and education in a short time would be needed in political decision-making.

Beyond considerations about the implementation of NETs, this chapter showed that the ethical challenge of dealing with climate change not only concerns the differentiation of responsibilities and entitlements with regard to global climate action as addressed in much of the climate ethics literature. Ethical challenges also occur once these kinds of questions have been settled. They concern the injustices established by implementing climate action. For the case of NETs, I argued that considerations taken from empirical environmental justice research may become relevant. This might also be true for other areas of climate action involving additional burdens and risks, such as mitigation, adaptation, or the governance of climate loss and damage. But beyond these kinds of considerations, other concerns of ethics and justice might become important. These involve questions about appropriate compensatory measures, the distribution of responsibilities between different national and subnational agencies for climate action and their empowerment, and even the fair distribution of the benefits from climate measures. Furthermore, this chapter showed why feasibility constraints to just climate action must lead us to frame ethical challenges differently and might lead to different ways of approaching climate change.

NOTES

1. I would like to thank the participants of the CDR workshop in fall 2020 at the University of Fribourg Environmental Sciences and Humanities Institute for their valuable comments and objections. My special thanks go to Dominic Lenzi and Simon Milligan for their close reading of earlier versions of this paper.

2. The all-affected principle has been introduced to define the proper decision-making body of a democracy. Goodin argues that it should include all those whose morally relevant interests are affected by the policy decisions of a democratic state. In our context it defines who should be included in the decision-making body for governing NET facilities. I discuss this principle in more detail in the next section.

REFERENCES

Anderson, Elizabeth. 1999. "What Is the Point of Equality?" *Ethics* 109(2): 287–337.
Anderson, Kevin, and Glen Peters. 2016. "The Trouble with Negative Emissions." *Science* 354(6309): 182–83.
Boysen, Lena R., Wolfgang Lucht, Dieter Gerten, Vera Heck, Timothy M. Lenton, and Hans Joachim Schellnhuber. 2017. "The Limits to Global-Warming Mitigation by Terrestrial Carbon Removal." *Earth's Future* 5(5): 463–74.

Brulle, Robert J., and David N. Pellow. 2006. "Environmental Justice: Human Health and Environmental Inequalities." *Annual review of public health* 27: 103–24.

Bui, Mai, Claire S. Adjiman, André Bardow, Edward J. Anthony, Andy Boston, Solomon Brown, Paul S. Fennell et al. 2018. "Carbon Capture and Storage (CCS): The Way Forward." *Energy & Environmental Science* 11(5): 1062–76.

Callies, Daniel Edward. 2018. "Institutional Legitimacy and Geoengineering Governance." *Ethics, Policy & Environment* 21(3): 324–40.

Caney, Simon. 2010. "Climate Change and the Duties of the Advantaged." *Critical Review of International Social and Political Philosophy* 13(1): 203–28.

Dooley, Kate, Ellycia Harrould-Kolieb, and Anita Talberg. 2020. "Carbon-dioxide Removal and Biodiversity: A Threat Identification Framework." *Glob Policy* 12: 34–44.

Downs, Anthony. 1957. "An Economic Theory of Political Action in a Democracy." *Journal of Political Economy* 65(2): 135–50.

Fogarty, John, and Michael McCally. 2010. "Health and Safety Risks of Carbon Capture and Storage." *JAMA* 303(1): 67–68.

Fourie, Carina, Fabian Schuppert, and Ivo Wallimann-Helmer, eds. 2015. *Social Equality: On What It Means to Be Equals*. Oxford: Oxford University Press.

Fraser, Nancy. 2010. *Scales of Justice: Reimagining Political Space in a Globalizing World*. Paperback ed. New Directions in Critical Theory. New York: Columbia Univ. Press.

Fuss, Sabine, J. G. Canadell, Glen Peters, Massimo Tavoni, Robbie M. Andrew, Philippe Ciais, Robert B. Jackson et al. 2014. "Betting on Negative Emissions." *Nature Climate Change* 4: 850–53.

Fuss, Sabine, William F. Lamb, Max W. Callaghan, Jérôme Hilaire, Felix Creutzig, Thorben Amann, Tim Beringer et al. 2018. "Negative Emissions—Part 2: Costs, Potentials and Side Effects." *Environ. Res. Lett.* 13(6): 63002.

Fyson, Claire L., Susanne Baur, Matthew Gidden, and Carl-Friedrich Schleussner. 2020. "Fair-Share Carbon Dioxide Removal Increases Major Emitter Responsibility." *Nat. Climate Change* 10(9): 836–41.

Gardiner, Stephen M. 2004. "Ethics and Global Climate Change." *Ethics* 114: 555–600.

Gardiner, Stephen M. 2020. "Ethics and Geoengineering: An Overview." In *Global Changes: Ethics, Politics and Environment in the Contemporary Technological World*. Vol. 46, edited by Luca Valera and Juan C. Castilla. 1st ed. 2020, 69–78. Ethics of Science and Technology Assessment 46. Cham: Springer International Publishing.

Gardiner, Stephen M., Simon Caney, Dale Jamieson, and Henry Shue, eds. 2010. *Climate Ethics: Essential Readings*. Oxford, New York: Oxford University Press.

Geden, Oliver. 2016. "The Paris Agreement and the Inherent Inconsistency of Climate Policymaking." *WIREs Climate Change* 7(6): 790–97.

Gilabert, Pablo, and Holly Lawford-Smith. 2012. "Political Feasibility: A Conceptual Exploration." *Political Studies* 60(4): 809–25.

Goodin, Robert E. 1989. "Theories of Compensation." *Oxford Journal of Legal Studies* 9(1): 56–75.

Goodin, Robert E. 2007. "Enfranchising All Affected Interests, and Its Alternatives." *Philosophy and Public Affairs* 35: 40–68.
Gosepath, Stefan. 2015. "The Principles and the Presumption of Equality." In *Social Equality: On What It Means to Be Equals,* edited by Carina Fourie, Fabian Schuppert, and Ivo Wallimann-Helmer 167–85. Oxford: Oxford University Press.
Gough, Clair, and Sarah Mander. 2019. "Beyond Social Acceptability: Applying Lessons from CCS Social Science to Support Deployment of BECCS." *Current Sustainable Renewable Energy Reports* 6(4): 116–23.
Hale, Benjamin. 2012. "The World That Would Have Been: Moral Hazard Arguments Against Geoengineering." In *Engineering the Climate: The Ethics of Solar Radiation Management*, edited by Christopher J. Preston, 113–32. Lanham: Lexington Books.
Hansson, Sven Ove. 2013. *The Ethics of Risk: Ethical Analysis in an Uncertain World.* New York: Palgrave Macmillan.
Hayward, Tim. 2012. "Climate Change and Ethics." *Nature Climate Change* 2(12): 843–48.
Honegger, Matthias, Steffen Münch, Annette Hirsch, Christoph Beuttler, Thomas Peter, Wil Burns, Oliver Geden et al. 2017. "Climate Change, Negative Emissions and Solar Radiation Management: It Is Time for an Open Societal Conversation." https://www.uni-heidelberg.de/md/hce/risk_dialogue_foundation_ce-dialogue_white_paper_17_05_05.pdf.
IPCC. 2018. "Global Warming of 1.5°C: An IPCC Special Report on the Impacts of Global Warming of 1.5°C Above Pre-Industrial Levels and Related Global Greenhouse Gas Emission Pathways, in the Context of Strengthening the Global Response to the Threat of Climate Change, Sustainable Development, and Efforts to Eradicate Poverty." Unpublished manuscript, last modified August 16, 2020. Summary for Policymakers.
Kortetmäki, Teea, and Markku Oksanen. 2016. "Food Systems and Climate Engineering: A Plate Full of Risks or Promises?" In *Climate Justice and Geoengineering: Ethics and Policy in the Atmospheric Anthropocene*, edited by Christopher J. Preston, 121–35. London: Rowman & Littlefield International Ltd.
Krütli, Pius. 2010. *Radioactive Waste Management: Justice and Decision-Making Processes in Repository Siting Environmental Sciences, ETH, Zurich.* Zürich: DISS. ETH NO. 19016.
Krütli, Pius, Kjell Törnblom, Ivo Wallimann-Helmer, and Michael Stauffacher. 2015. "Distributive Versus Procedural Justice in Nuclear Waste Repository Siting." In *The Ethics of Nuclear Energy: Risk, Justice and Democracy in the Post-Fukushima Era*, edited by Behnam Taebi and Sabine Roeser, 119–40. Cambridge: Cambridge University Press.
Lenzi, Dominic. 2018. "The Ethics of Negative Emissions." *Global Sustainable* 1: e7.
Lenzi, Dominic, William F. Lamb, and Jérôme Hilaire, Jérôme. 2018. "Weigh the Ethics of Plans to Mop up Carbon Dioxide." *Nature* 561: 303–305.
Lin, Albert C. 2013. "Does Geoengineering Present a Moral Hazard?" *Ecology Law Quarterly* 40(3): 673–712. Accessed December 23, 2020.

Minx, Jan C., William F. Lamb, Max W. Callaghan, Lutz Bornmann, and Sabine Fuss. 2017. "Fast Growing Research on Negative Emissions." *Environmental Research Letters* 12(3): 35007.
Minx, Jan C., William F. Lamb, Max W. Callaghan, Sabine Fuss, Jérôme Hilaire, Felix Creutzig, Thorben Amann et al. 2018. "Negative Emissions—Part 1: Research Landscape and Synthesis." *Environment Research Letters* 13(6): 63001.
Morello-Frosch, Rachel, Miriam Zuk, Michael Jerrett, Bhavna Shamasunder, and Amy D. Kyle. 2011. "Understanding the Cumulative Impacts of Inequalities in Environmental Health: Implications for Policy." *Health Affairs (Project Hope)* 30(5): 879–87.
Niemeyer, Simon. 2014. "A Defence of (Deliberative) Democracy in the Anthropocene." *Ethical Perspectives* 21(1): 15–45.
O'Neill, John. 2017. "The Price of an Apology: Justice, Compensation and Rectification." *Cambridge Journal of Economics* 41(4): 1043–59.
Page, Edward A. 2012. "Give It up for Climate Change: A Defence of the Beneficiary Pays Principle." *Int. Theory* 4(2): 300–30.
Pamplany, Augustine, Bert Gordijn, and Patrick Brereton. 2020. "The Ethics of Geoengineering: A Literature Review." *Science and Engineering Ethics* 26(6): 3069–119.
Pastor, Manuel, Jim Sadd, and John Hipp. 2001. "Which Came First? Toxic Facilities, Minority Move-In, and Environmental Justice." *Journal of Urban Affairs* 23(1): 1–21. Accessed October 22, 2017.
Pasztor, Janos. 2017. "The Need for Governance of Climate Geoengineering." *Ethics International Affair* 31(4): 419–30.
Pozo, Carlos, Ángel Galán-Martín, David M. Reiner, Niall Mac Dowell, and Gonzalo Guillén-Gosálbez. 2020. "Equity in Allocating Carbon Dioxide Removal Quotas." *Nature Climate Change* 10(7): 640–46.
Roser, Dominic. 2016. "Reducing Injustice Within the Bounds of Motivation." In *Climate Justice in a Non-Ideal World*, edited by Jennifer C. Heyward and Dominic Roser. First edition, 83–103. Oxford: Oxford University Press.
Scheffler, Samuel. 2015. "The Practice of Equality." In *Social Equality: On What It Means to Be Equals*, edited by Carina Fourie, Fabian Schuppert, and Ivo Wallimann-Helmer, 20–44. Oxford: Oxford University Press.
Schlosberg, David. 2007. *Defining Environmental Justice: Theories, Movements, and Nature*. 1. publ. Oxford u.a. Oxford Univ. Press.
Schuppert, Fabian, and Ivo Wallimann-Helmer. 2014. "Environmental Inequalities and Democratic Citizenship: Linking Normative Theory with Empirical Research." *Analyse & Kritik* 36(2): 345–66.
Shue, Henry. 1999. "Global Environment and International Inequality." *International Affairs* 75(3): 531–45.
Southwood, Nicholas. 2018. "The Feasibility Issue." *Philosophy Compass* 13(8): e12509.
Steigleder, Klaus. 2017. "The Tasks of Climate Releated Energy Ethics—The Example of Carbon Capture and Storage." *Jahrbuch für Wissenschaft und Ethik* 21: 121–46. Accessed March 02, 2021.

UNFCCC (United Nations Framework Convention on Climate Change). 2015. "Adoption of the Paris Agreement." Decision 1/CP.21. Document FCCC/CP/2015/10/Add.1. Paris.

Valentini, Laura. 2012. "Justice, Disagreement, and Democracy." *British Journal of Political Science* 43(1): 177–99.

van Parijs, Philippe. 2011. *Just Democracy: The Rawls-Machiavelli Programme*. ECPR press essays. Colchester: ECPR Press.

Walker, Gordon. 2010. "Environmental Justice, Impact Assessment and the Politics of Knowledge: The Implications of Assessing the Social Distribution of Environmental Outcomes." *Environmental Impact Assessment Review* 30(5): 312–18.

Walker, Gordon. 2012. *Environmental Justice: Concepts, Evidence and Politics*. London u.a. Routledge.

Wallimann-Helmer, Ivo. 2015. "Justice for Climate Loss and Damage." *Climatic Change* 133(3): 469–80.

Wallimann-Helmer, Ivo. 2019a. "Justice in Managing Global Climate Change." In *Managing Global Warming: An Interface of Technology and Human Issues*, edited by Trevor Letcher, 751–68. London: Cambridge Academic Press.

Wallimann-Helmer, Ivo. 2019b. "The Ethics of Waste Policy." In *The Routledge Handbook of Ethics and Public Policy*, edited by Annabelle Lever and Andrei Poama, 501–12. Routledge Handbooks in Applied Ethics. New York: Taylor & Francis.

Wallimann-Helmer, Ivo, Lukas Meyer, and Paul Burger. 2016. "Democracy for the Future: A Conceptual Framework to Assess Institutional Reform." *Jahrbuch für Wissenschaft und Ethik* 21: 197–222.

Wiens, David. 2015. "Political Ideals and the Feasibility Frontier." *Economics and Philosophy* 31(3): 447–77.

Williamson, Phil. 2016. "Scrutinize CO2 Removal Methods." *Nature* 530: 153–55.

Zhang, Zhihua, and Donald Huisingh. 2017. "Carbon Dioxide Storage Schemes: Technology, Assessment and Deployment." *Journal of Cleaner Production* 142: 1055–64.

Index

ability-to-pay principle (APP), 4, 10, 62, 72, 80, 198. *See also* preventative APP
accessibility, 42, 173
Action Agenda for Nature and People, 48
action guidance, 7
activists, 75, 153
adaptation, 3, 38, 42, 47, 48, 50, 51, 71, 93, 105, 106, 127, 130, 150–52, 161, 164
adaptation finance, 78, 158, 159
Adaptation Gap Report, 71
Ad Hoc Working Group on the Paris Agreement (APA), 153
afforestation, 193, 195, 197
AFOLU. *See* agriculture, and other land-use changes to capture carbon
Agenda 2030, 35, 38, 42
Agenda for Sustainable Development (2030), 43, 50. *See also* Agenda 2030
agents, 6, 8, 19, 63, 64, 97, 99, 124, 132, 134, 172. *See also* responsible agents
agriculture, 26, 42, 117, 130, 193, 195
agriculture, and other land-use changes to capture carbon (AFOLU), 193, 194

all-affected principle, 199
Alliance of Small Island States (AOSIS), 76, 78, 151, 178, 180, 182, 184–87
ambition gap, 67–70, 78–80
American Society of International law, 104
AOSIS. *See* Alliance of Small Island States
APP. *See* ability to pay principle
AR5. *See* Fifth Assessment Report
argument from individual duties, 97
attribution, 155, 156
Australia, 77

Bangladesh, 180, 181, 183, 184
BASIC countries, 69, 75, 76, 78
basic rights, 133, 142, 202
BECCS. *See* bioenergy production with carbon capture and storage
beneficiary pays principle (BPP), 4, 10, 62, 101, 196, 197
Biden, Joe, 7, 81n11
binary approach, 36
biodiversity, 37, 45, 50, 194
bioenergy production with carbon capture and storage (BECCS), 22, 193, 194, 196, 197
biofuels, 24, 26

biomass, 193, 194, 196–98
Board of the Green Climate Fund (GCF), 152
BPP. *See* beneficiary pays principle
Brazil, 69
Brazil Proposal, 4
Brennan, Geoffrey, 135
Broome, John, 8
Brussels Supplementary Convention on Third Party Liability in the field of Nuclear Energy, 151
Buchanan, Allen, 173
burden-sharing justice, 59

CAC. *See* contraction and convergence approach
Cancun Adaptation Framework, 151
Caney, Simon, 7, 8, 19, 77, 94, 98
capabilities approach, 49, 50
carbon budget, 60, 64, 115
carbon capture, 18, 191–93
carbon dioxide (CO_2), 18, 160, 191
carbon dioxide removal (CDR), 18, 22, 25
carbon emissions, 121, 134
carbon-intensive economy, 25
carbon storage, 136, 195
Caribbean Catastrophe Risk Insurance Facility, 154
CBD. *See* Convention of Biological Diversity
CBDR-RC. *See* common but differentiated responsibilities and respective capabilities
CDR. *See* carbon dioxide removal
central governments, 38
CERF. *See* Climate equity reference framework
CH_4 (methane), 160
China, 75, 78, 105
circumstances of justice, 6
civil society, 41, 51, 60, 74–77, 80
clean energy, 42
climate action, 2, 8, 10, 41, 47, 48, 89, 98, 152, 157, 163, 172, 191, 196, 197, 206

Climate Action Tracker, 77
climate adaptation finance, 78
climate change, 1, 4, 6, 9, 18, 24, 35, 37, 40, 44–46, 51, 59, 63, 67, 74, 80, 89, 95, 98, 135, 156, 171, 173, 191; and APP, 92, 93, 100, 101, 103, 106; dangerous, 68, 133, 137, 192; and emissions, 125–27, 130, 150, 159, 162, 196; global, 3, 5, 7, 160; harm/impact, 42, 105, 154; loss and damage from, 149, 151, 156, 157, 160–62, 164, 165; man-made, 118–20; and mobility, 173, 175–78, 180–86
Climate Change and Land report, 42
climate change governance, 50
climate change policy, 39, 50
climate change regime, 37
climate crisis, 159
climate displacement, 175
climate economics, 22
climate equity reference calculator, 70, 74
climate equity reference framework (CERF), 70–74
climate ethicists, 93, 172
climate ethics, 92, 171
climate extortion, 9
climate finance, 38, 70, 105, 149, 157, 158, 164
climate goals, 39, 101, 105
climate governance, 24, 41, 44, 46, 48
climate harm, 3
climate insurance, 155
climate justice, 3, 5, 7–10, 17, 18, 20, 21, 24, 27, 28, 36, 37, 89–91, 103, 107, 191, 196
climate justice index, 69, 77, 79
climate justice theorists, 7
climate laggards, 75, 76, 80
climate mitigation policy pathways, 23
climate models, 10, 191–94, 200, 203, 205
climate policy, 4, 5, 15, 17, 42, 45, 59, 60, 68, 76, 78, 80, 132
climate policy community, 25

Index

climate policy pathways, 16, 28
climate regime, 39, 43, 47, 50, 51, 60, 67, 68, 71, 73, 76, 79, 80, 151
climate scenarios, 20
climate science, 15, 17, 18, 21
climate scientists, 16, 119
climate skeptics, 26
climate treaty, 8, 9, 19, 65, 66
CO_2. *See* carbon dioxide
cognitive values, 21
collaborative role, 22, 24
common but differentiated responsibilities and respective capabilities (CBDR-RC), 4, 60, 68–74, 78–80, 102
compatriots, 94, 96, 97
compensation, 60, 105, 106, 120, 131, 139, 155, 156, 159, 163, 164, 198, 200, 201
compensatory justice, 160, 162, 198
Conference of the Parties (COP), 2, 67, 74, 151, 152, 181, 191
conservatives, 105
constructivist international relations theory, 68, 80
contraction and convergence approach (CAC), 133–37
Convention of Biological Diversity (CBD), 40, 44, 47
COP. *See* Conference of the Parties
coronavirus, 64. *See also* COVID-19
cosmopolitan theory, 61
Costa Rica, 178
COVID-19, 7, 162. *See also* pandemic
Crawford, Neta, 68, 78
Crocker, David, 49

DACCS. *See* direct air carbon capture and storage
decarbonization, 19, 25, 92, 94
decision-making, 24, 25, 36, 37, 104, 199, 202–6
deliberation process, 17, 24–27
democratic citizenship, 202, 205
democratic legitimacy, 24

Democratic Party, 7
descriptive (norms), 16, 18, 19, 21
desirability, 9, 16, 20, 22, 24, 28, 109
developed countries, 4, 6, 62, 65, 70, 74, 89, 100, 149, 151, 152, 156, 158–60, 197
developing countries, 2, 44, 72, 78, 117, 120, 122, 123, 135, 140, 149, 151, 152, 156, 159, 163
development assistance, 97, 164
development threshold, 69–71, 73, 74
Dewey, John, 23, 27
diagnostic approach, 20
diagnostic role, 20, 24, 28
direct air carbon capture and storage (DACCS), 193, 196, 197
directional leadership, 176, 180, 181, 184
discounting approach, 49, 98
discursive process, 46, 50
displacement, 35, 175, 176
distributing emissions, 117
distributive justice, 1, 4, 62, 121, 122, 126, 128, 133, 135, 138, 139, 195, 197, 198
Dryzek, John, 41
Durban Platform, 102

EcoEquity, 74
economic efficiency, 15
economic rationality, 16
efficiency, 194, 198, 201, 203, 205
efficiency blinders, 45
efficiency without sacrifice model, 8
embedded proceduralism, 10, 35, 46, 47, 51
emerging countries, 62, 76, 78, 117, 121, 197. *See also* developing countries
emissions, 3, 4, 6, 8, 16, 18, 26, 43, 65, 66, 71, 73, 80, 92, 115, 124–26, 137, 150, 162, 196; budget, 116, 119, 121, 122, 129, 132, 136, 138; negative, 191–93, 197; reductions, 1, 70, 102, 131, 132, 135, 144; rights, 117, 120, 123, 130, 131, 133

Emissions Gap Report, 71, 76
emissions-generating activities, 116–23, 127, 128, 131, 138
energy transition, 79
engagement (moral), 21, 38, 40, 41, 44
entitlement, 196, 206
environmental justice, 195, 196, 198
environmental risks, 195–99, 201, 205
EPC. *See* equal per capita approach
epistemic quality, 25
epistemic values, 21, 22
equality, 24
equal per capita approach (EPC), 117, 119–21, 126, 127, 129, 133, 134, 136, 137
equal-per-capita principle, 196
equitable cooperation, 4
equity, 4, 10, 17, 18, 20, 35, 67, 68, 102
equity band, 70–74, 80, 82
equity calculator, 60, 70–80
equity gap, 75, 76
Equity Gap Report, 71, 76
Equity Review Coalition, 75, 76
EU (European Union), 6, 64, 70, 78
ExCom (Executive Committee of Conference of Parties), 151–54, 157
ex post criticism, 17, 20
extortion, 102, 137
extrication ethics, 63. *See also* transition ethics
ExxonMobil, 160

fact-value entanglement, 17, 22, 28
Falkner, Robert, 38
feasibility, 6, 8, 10, 18, 21, 23, 28, 60, 64, 68, 74, 77, 96, 136, 140, 149, 154, 157, 162, 172, 174; concerns, 123, 132, 173; judgments, 16, 25, 26; political, 23, 36, 44, 89–93, 95, 99, 100, 104, 105; technical, 154, 155, 163, 165
feasibility assessments, 15, 17, 20, 24, 172, 173
feasibility constraints, 5, 11, 39, 46, 60, 64, 79, 116, 132, 136, 138, 140, 191–95, 197, 202, 206

feasible set, 20–22, 60, 64–66, 76, 79
Fifth Assessment Report (AR5), 4, 17, 23, 26, 27, 67
Fiji, 46
First Assessment Report, 67
first order responsibilities, 8
followers, 172, 177, 186
food security, 194
formal requirement of citizenship, 202
fossil fuel companies, 160, 161, 163
fossil fuel lobby, 75, 78, 80
fossil fuel providers, 68, 79
fossil fuels, 2, 63
Framework Convention, 102, 105
free-ride states, 99
French government, 163
future generations, 20, 22, 67, 203–5

Galston, William, 16
Gardiner, Stephen, 5, 9, 19, 39, 63, 102
GCB. *See* global carbon budget
GCC. *See* Gulf Cooperation Council
GCF. *See* Green Climate Fund
GDR. *See* Greenhouse development rights
Germany, 178
GHG. *See* greenhouse gases
Gilabert, Pablo, 173, 174, 187
global biodiversity crisis, 35
global carbon budget (GCB), 60, 64, 79, 115, 117, 120–22, 126, 132–36
Global Climate Action Agenda, 41
Global Climate Action portal, 41
global climate crisis, 35
global climate governance, 41, 42, 44, 48, 51
global climate regime, 37
global governance, 40, 52, 63
globalism, 100
global justice, 89, 91. *See also* international justice
global mean temperature, 191, 192, 205
global poor, 89, 101
global temperature rise, 48. *See also* global warming
global warming, 1–3, 37, 60, 192

Index

GNI. *See* Gross National Income
GNP (gross national product), 100
Goodin, Robert E., 200
governance by goal setting, 43
grandfathering, 116, 127, 129, 132–35
Green Climate Fund (GCF), 152, 157
green energy, 16
Greenhouse development rights (GDR), 69, 70, 73
greenhouse gases (GHG), 1–3, 5, 8, 26, 43, 64, 71, 79, 93, 117, 119, 150, 156, 157, 159, 160
Greenpeace, 16
Griffin, Paul, 160
gross domestic product (GDP), 98
gross national income, 93, 100
Gulf Cooperation Council (GCC), 75, 76, 80

Hale, Thomas, 38
Hansen, James, 64
hard constraints, 36, 174
harm avoidance justice, 59
health risks, 201
Henkin, Louis, 104
Heyward, Clare, 19
high-emitting countries, 126, 130, 132, 135, 139, 144
high-emitting nations, 19. *See also* high-emitting countries
high-emitting states, 19, 65, 116, 129–32, 137–40
historical emissions, 20, 72, 101, 115, 116, 118, 119, 121–23, 127, 128, 132–39, 160. *See also* past emissions
historical responsibility, 59, 70, 72, 123, 134, 135, 137, 140
Human Development Index (HDI), 74
human flourishing, 10, 35–37, 49–51
human mobility, 178, 180, 181
human rights violations, 98

IAPAL. *See* International Air Passenger Levy
idea-based leadership, 176, 177, 180–84

ideals (moral), 21, 91
ideal theory, 6, 59–62, 76
illegitimate expectations, 124, 125
Indonesia, 78
industrialization, 117, 121, 122
industrialized countries, 73, 116, 117, 120, 123, 132, 133, 136, 157, 164
industrialized nations, 4, 151. *See also* industrialized countries
industrialized states, 106, 130, 133–37, 139
inequality, 24, 97, 120, 128, 138
Inside Climate News, 160
institutional reform, 10, 60, 67, 68, 72, 73, 75, 76, 204
instrumental leadership, 177, 180–85
insurance, 154, 155, 159, 194
integrated assessment models, 16
Interagency Liaison Group (ILG), 45
interdependence model, 22, 23, 46
intergenerational cooperation, 4
intergenerational equity, 18, 22
intergenerational justice, 89, 126, 127, 131, 132, 138
Intergovernmental Panel on Climate Change (IPCC), 3, 9, 15, 16, 18, 23, 24, 26, 28, 42, 60, 67, 71, 74, 77, 118, 160, 193
International Air Passenger Levy (IAPAL), 161, 162
International Air Passenger Loss and Damage Levy (IAPLDL), 162
international climate law, 101–3
international climate regime, 60, 77, 94, 151
International Development Strategy, 100
international fairness, 91. *See also* international justice
international insurance pool, 151
international justice, 6, 126, 127, 131, 138, 139
international law, 66, 68, 104, 105, 157
international mitigation cooperation, 70
international negotiations, 1, 60, 64, 67, 79, 80, 105, 149, 173, 176

International Oil Pollution Compensation Fund (IOPCF), 163
International Organization for Migration (IOM), 178
international paretianism, 8, 19, 65
intuitions (moral), 93, 95, 96, 98, 103
IPCC. *See* Intergovernmental Panel on Climate Change
Ireland, 26

Jamieson, Dale, 5
Japan, 70
judgement (moral), 71
jus cogens norms, 96
justice, 6, 7, 9, 10, 175
justice-improvement, 6
justificatory role, 20, 24, 28

Keohane, Robert, 39, 104
Kingston, Ewan, 74, 75, 77
Klein, Naomi, 75
Kyoto Protocol, 2, 4

Lawford-Smith, Holly, 172–74, 186
LDC. *See* least developed countries
leadership deficit, 174, 185, 187
leadership style, 175, 177, 183–85
least developed countries (LDC), 20, 69, 150, 157, 161, 162
legitimate expectations, 123–29, 131, 132
levy, 10, 94, 161–65. *See also* taxes
lexical priority, 93, 99, 101, 102
liability, 159, 160, 162, 163
liability and compensation, 106, 150, 153, 154, 156, 157
liberal theory, 61
Light, Andrew, 72, 78
liquified carbon, 198
Longino, Helen, 21
loss and damage, 105–7
loss and damage finance, 149, 153, 157–59, 161, 164, 165
Low, Sean, 25

low-carbon economy, 18

Maljean-Dubois, Sandrine, 38
Marrakech Partnership for Global Climate Action, 41
McKibben, Bill, 75
Michaelowa, Axel, 78
Michaelowa, Katharina, 78
migration, 175, 176, 180
mitigation, 15–18, 20, 22, 45, 47, 49, 62, 67, 70–72, 76–79, 102, 105, 150, 152, 153, 159
mobility, 176
modeling injustice, 20
modes of leadership, 176, 179
Moellendorf, Darrel, 104
moral agents, 62, 96
moral hazard argument, 194
morally relevant affectedness, 204
moral motivations, 78
moral norms, 95
moral psychology, 5
moral urgency, 7
Morocco, 178
motivation, 20, 156, 194
Muller, Benito, 161

Nansen Initiative, 178, 181
nationally determined contributions (NDC), 3, 10, 48–51, 60, 69–71, 73, 75–80, 93, 116
NDC. *See* nationally determined contributions
negative emission technologies (NETs), 11, 191–206
Negishi weights, 23
NET. *See* negative emission technologies
net zero emissions, 126, 128, 129, 132, 139
new realism, 65, 66
NGO. *See* nongovernmental organizations
Niemeyer, Simon, 26

Noel, Alain, 97
noncompliance, 6, 8, 62, 63
noneconomic loss and damage (NELD), 155
nongovernmental organizations (NGO), 41, 74, 75, 77
nonideal approach, 60, 63
nonideal justice, 21, 62
nonideal theory, 6, 21, 60–63, 79
nonidentity problem, 118, 122, 141n7
non-state actors, 38, 40–42, 48, 49, 178
normative assessment, 124
normative claims, 125
normative dimension, 49, 74
normative discourse, 37
normative elements, 20
normative expertise, 24
normative goals, 66
normative issues, 18, 71
normative literature, 4, 18, 52
normative reasons, 20, 116
normative research, 18, 21
normative theory, 15
Norway, 178, 180, 181, 183–85, 187
nuclear proliferation, 176
nuclear waste, 199

objection from ephemerality, 103
objection from irrelevance, 104
Ohlin, Jens, 66, 68
opinion surveys, 16
outcomes, 45, 68, 77, 78, 94, 95, 97, 173, 174, 185, 186, 192
Oxfam, 74

pandemic, 7, 64, 105, 162
Paprocki, Kasia, 164
Paris Agreement, 1–4, 10, 37–40, 48, 50, 77, 101, 102, 115, 129, 153, 158; and loss and damage, 106, 153, 158; and NDCs, 60, 68, 71, 73, 76, 80, 116; and SDGs, 41–43, 49; target goals, 15, 192–94; and United States, 7, 103, 104, 152

past emissions, 10, 59, 72, 73, 89, 101, 115–23, 128, 132–34, 138
paternalism, 201
Pearson Commission, 100
philosophers, 17, 20, 22, 23, 27, 36, 60, 65, 89, 91, 98, 171
philosophy of science, 21
planned relocation, 176
pluralism, 47
Poland, 46
polarization, 26
policymakers, 4, 25, 66, 89, 91, 93, 95, 98, 105–7
policymaking, 1, 3, 4, 68, 80, 91, 107, 150, 152, 162, 202, 204
policy preferences, 16, 25
political constraints, 64, 79
political dialogue, 35–37, 41, 46–49, 51
political philosophy, 6, 21, 61, 63, 76, 90
political theory, 90
polluter pays principle (PPP), 4, 10, 62, 66, 72, 80, 93, 99, 101, 103, 106, 149, 159, 160, 163, 165, 196, 197
pollution, 159, 161, 163–65
Posner, Eric, 19, 65, 66, 76, 118
poverty, 2, 50, 74
power imbalances, 164
power inequities, 25
PPP. *See* polluter pays principle
practical feasibility, 44
pragmatism, 23
pragmatist philosophy, 27
preventative APP, 92, 95
principle of equality, 196
Principle of Human Flourishing, 49
principle of reciprocity, 49, 50
principles of justice, 6. *See also* ideal theory
proceduralism, 47, 48, 51
procedural justice, 198, 199
progressive taxation, 95
promissory obligation, 109n29
public health, 7

Räikkä, Juha, 90
Rawls, John, 6, 21, 61, 62
reactionary citizens, 89, 91, 94–96, 98, 100, 104–7
realism, 74, 95
realistic scenarios, 16
realistic utopia, 6
reciprocity, 51
redistribution of wealth, 23
reflexive learning process, 24
renewable energy, 16, 79, 136, 161
reparations, 162
reputational concerns, 60, 73, 76, 78
reputational costs, 77
responsible agents, 4, 8, 140
Rio Declaration, 159
Rio Earth Summit, 2, 38. *See also* United Nations Conference on Environment and Development
risk management, 152, 153
risk reduction, 194
Robeyns, Ingrid, 49
Roser, Dominic, 19
Rudner, Richard, 21
Russia, 75

Santiago Network, 152
Saudi Arabia, 75
Sayre-McCord, Geoffrey, 135
scalar approach, 36
Schäfer, Stefan, 25
Schuppert, Fabian, 19
SDG. *See* Sustainable Development Goals
second-order responsibilities, 8
Seidel, Christian, 19
self-interest, 73, 78, 79, 94, 100, 107, 115
shared socioeconomic pathways (SSP), 23
shared values, 10
Shue, Henry, 62, 63, 171
SIDS. *See* Small Island Developing States
Simmons, John, 21

Singapore, 78
skepticism, 28, 66
Small Island Developing States (SIDS), 150, 182, 184, 186
social justice, 26
sociocultural feasibility, 17
socioeconomically disadvantaged, 192, 197, 199
soft feasibility constraints, 36, 174, 186, 187
Solar PV, 18
solar radiation, 192, 193
Solidarity Levy, 163
South Africa, 78
sovereign states, 6
Special Report on Global Warming of 1.5°C, 3, 9, 15, 17, 18, 24, 67
Special Report on Renewables, 16
Special Report on the Ocean and Cryosphere, 67
sticky institutional structures, 6
Stockholm Environment Institute, 74
structural leadership, 176, 180, 181, 184–87
subnational actors, 38
substantial requirement of citizenship, 202, 203
surface temperature, 2
Sustainable Development Goals (SDG), 35, 37, 38, 41–44, 46, 49, 50, 52, 97
Suva Expert Dialogue on Loss and Damage, 154, 158
Swift, Adam, 20, 21
Switzerland, 178, 199
synergies, 37, 38, 152

Talanoa Dialogue, 46
taxes, 26, 94, 95, 158, 163; on carbon, 16, 162; on polluters, 10, 160, 161, 164
Technical Expert Group on Comprehensive Risk Management (TEG-CRM), 154
theoretical feasibility, 90
Therein, Jean-Phillipe, 97

thick ethical concepts, 22
threat multiplier, 45
350.org, 74
Thunberg, Greta, 75
transitional improvements, 72, 79
transitional principle, 73
transitional standpoints, 187
transition ethics, 63
Trump, Donald, 104

UN. *See* United Nations
Underdal, Arild, 177, 185
unequal benefits, 118–20, 122, 123
UNFCCC. *See* United Nations Framework Convention on Climate Change
Unitaid (health initiative), 163
United Nations (UN), 38, 100, 149, 156, 181
United Nations Climate Change Conference, 175
United Nations Conference on Environment and Development (UNCED), 38
United Nations Environment Programme (UNEP), 76
United Nations Framework Convention on Climate Change (UNFCCC), 4, 9, 46, 52, 71, 152; and the Paris Agreement, 73, 102, 104; party agreements under, 67, 73, 76, 78, 79; process of, 2, 10, 43
United Nations High Commissioner for Refugees (UNHCR), 178
United States, 2, 7, 65, 70, 75–77, 79, 100, 103, 104, 118, 157
U.S. Senate, 7

utilitarian theory, 61
utopian theory, 107

Valentini, Laura, 61
value-action gap, 27
value-free ideal, 21
values, 17, 20–23, 26, 38, 40, 71, 80, 150, 155; of reciprocity, 35–37, 50; substantive, 38, 42, 46–48, 51
Vanderheiden, Steve, 59
Vanuatu, 151
violations (moral), 93, 104
voluntarist APP, 89, 90, 92, 99–107

warming target (2°C), 23
Warsaw International Mechanism on Loss and Damage (WIM), 151–54, 157
waste policy, 195
wealth disparities, 18
wealth redistribution, 10, 16, 164, 165
wealthy countries, 97, 101, 102, 150, 157, 164
wealthy nations, 161. *See also* wealthy countries
wealthy states, 96, 103, 106
Weisbach, David, 19, 39, 65, 66, 76, 118
WIM. *See* Warsaw International Mechanism on Loss and Damage
wrongful harm, 5, 128
WWF (World Wildlife Fund), 74

Young, Iris Marion, 185

zero emissions. *See* net zero emissions

About the Contributors

Idil Boran is an associate professor in the Department of Philosophy (Faculty of Liberal Arts and Professional Studies) and faculty member at the Dahdaleh Institute for Global Health Research, at York University. She holds a dual appointment to the Graduate Program in Philosophy and the Graduate Program in Environmental Studies at York University, alongside an external affiliation as an associate researcher (nonresident) with the German Development Institute/Deutsches Institut für Entwicklungspolitik (DIE) located in Bonn, Germany. She is the founder and lead researcher of the Synergies of Planetary Health Research Initiative & Lab, an international and transdisciplinary collaborative initiative for research and global outreach, housed at the Dahdaleh Institute for Global Health Research. Author of *Political Theory and Global Climate Action: Recasting the Public Sphere* (Routledge 2019), Boran has years of experience with research and outreach activities in the UN Climate Change (UNFCCC) sphere.

Michel Bourban is a postdoctoral researcher in politics and international studies at the University of Warwick (UK). He is working on a project titled "Cosmopolitan Citizenship and Ecological Citizenship," funded by the Swiss National Science Foundation, under the supervision of Simon Caney. He completed his Ph.D. in philosophy at the University of Lausanne (Switzerland) and the University of Paris-Sorbonne (France), and held a postdoctoral researcher and lecturer position in philosophy and ethics of the environment at Kiel University (Germany). His work on climate justice, climate ethics, ecological citizenship, effective altruism, and innovation ethics has been published in journals such as *Ethics, Policy & Environment*, *Philosophy & Technology*, and *De Ethica*, and in edited volumes published by Routledge, Springer, Elsevier, and Göttingen University Press. He has also published two books on climate justice and climate ethics with the French publishers PUF and Vrin.

Md Fahad Hossain is a research officer at the International Centre for Climate Change and Development (ICCCAD), where he oversees the implementation of the project "Supporting Least Developed Countries (LDCs) on Loss and Damage." He also works with the LDC Universities' Consortium on Climate Change (LUCCC), an official capacity-building initiative of LDCs hosted by ICCCAD. His research interests lie in the area of climate change adaptation and loss and damage, with a specific focus on public policy. Fahad received his BSS in economics from the University of Dhaka, and MSc in climate change and development from Independent University, Bangladesh.

Saleemul Huq is the director of the International Centre for Climate Change and Development (ICCCAD), and a senior fellow at the International Institute for Environment and Development (IIED). He also holds a senior advisor position at the Global Centre on Adaptation (GCA). Saleemul's recent work focuses on building negotiating capacity and supporting the engagement of the least developed countries (LDCs) in the UNFCCC, including work on negotiator training workshops for LDCs, policy development, and research into vulnerability and adaptation to climate change in the LDCs. He has published over 100 articles in scientific journals. He was a lead author of the Intergovernmental Panel on Climate Change's (IPCC) Third Assessment Report, and one of the coordinating lead authors of its Fourth Assessment Report. In 2019, *Apolitical* named him in the top 20 among the "World's 100 Most Influential People in Climate Policy."

Danielle Falzon is a Ph.D. candidate in sociology at Brown University, USA. Her work focuses on power and inequality in decision-making on climate change. Her initial work in this area examined the institutional shortcomings of the UN climate negotiations, both with regard to how the institution is structured to privilege the capacities of wealthy, polluting countries and how the negotiations themselves are ill-suited to generate an effective climate treaty. Her dissertation work delves into climate adaptation in Bangladesh, examining the adaptation regime, exploring, in particular, how and why they approach adaptation as a business investment, and identifying the effects of this approach.

Corey Katz is an assistant professor of philosophy at Georgian Court University in Lakewood, NJ. He was a postdoctoral researcher in the ethics of sustainable development at the Ohio State University from 2016 to 2018 and received his Ph.D. from Saint Louis University. He has published on how we should understand our moral responsibilities to future generations, and on the ethics of greening healthcare, with papers appearing in *Ethical Theory and Moral Practice*, *Ethics, Policy & Environment*, and the *Journal of Medicine and Philosophy*. His research interests also include Kantian contractualism and the question of individual responsibility for overdetermined harms like climate change.

Sarah Kenehan is an associate professor of philosophy in the Department of Philosophy and Religious Studies at Marywood University in Scranton, PA. She earned her Ph.D. in 2010 from Graz University in Austria, with specializations in liberal political philosophy (esp. Rawls), international justice, and climate justice. She currently teaches and writes in various areas of ethics and social and political philosophy, including climate justice, Rawls, and animal ethics. Most recently, she coedited (with Erinn Gilson) and contributed to the volume *Food, Environment, and Climate Change: Justice at the Intersections* (RLI, 2019). Other recent work has appeared in *Climate Justice and Historical Emissions* (eds. L. Meyer and P. Sanklecha, Cambridge, 2017) and *The Ethics of Animal Experimentation: Working Towards a Paradigm Change* (eds. K. Herrmann and K. Jayne, Brill, 2018).

Ewan Kingston is trained as a philosopher, with a masters from Victoria University of Wellington, and a Ph.D. from Duke University. He is currently a Climate Futures Initiative postdoctoral researcher at Princeton, jointly in the Princeton Environmental Institute and the University Center for Human Values. His research has two strands. The first strand is business ethics—particularly examining how consumers, firms, and NGOs should respond to moral flaws in global supply chains. The second strand is climate ethics, where he has looked at the duties of states, individuals, and NGOs with regard to the climate crisis. In future research, he plans to weave these strands together to create an interdisciplinary, nonideal account of climate change obligations in the domain of business.

Martin Kowarsch heads the working group "Scientific Assessments, Ethics, and Public Policy" at the Mercator Research Institute on Global Commons and Climate Change (MCC). With a background in philosophy and economics, he led the joint research initiative of MCC and UN Environment on "The Future of Global Environmental Assessment Making" (2013–2017) to inform future choices in solution-oriented assessment design. With Ottmar Edenhofer, he developed the normative "Pragmatic-Enlightened Model" for assessment making, which influenced the assessment strategy of IPCC Working Group III (AR5).

Dominic Lenzi is a research associate at the Mercator Research Institute on Global Commons and Climate Change, in the working group "Scientific Assessments, Ethics, and Public Policy." His research focuses on climate justice and the ethics of geoengineering, political philosophy and natural resource justice, and environmental philosophy. Dominic is also a lead author and fellow of the Intergovernmental Science-Policy Platform on Biodiversity and Ecosystem Services (IPBES) Assessment of the Diverse Values of Nature.

Univ.-Prof. Dr. Lukas H. Meyer, M.A., Dipl.Pol. is professor of philosophy at the University of Graz, Austria, and speaker of the Working Unit Moral and Political Philosophy. He received his doctorate from Oxford University and completed his habilitation at the University of Bremen, was faculty fellow in ethics at Harvard University (2000–2001), Humboldt Research Fellow at Columbia University (2001–2002), and assistant professor of philosophy at the University of Berne (2005–2009). Since 2014 Meyer has been the speaker of the interdisciplinary Doctoral Program Climate Change (funded by the Austrian Science Fund), and since 2019, the speaker of the interfaculty Field of Excellence Climate Change Graz. Meyer served as Dean of the Faculty of Arts and Humanities at Graz from 2013 to 2017. As one of the first philosophers to work in the organization, Meyer served as a lead author for the Intergovernmental Panel on Climate Change (IPCC). His book publications include *Climate Change and Historical Emissions* (coedited, CUP 2017); *Intergenerational Justice* (coedited, OUP 2012); *Legitimacy, Justice and Public International Law* (edited, CUP 2009); and *Historische Gerechtigkeit (Historical Justice)* (de Gruyter 2005). Meyer is a founding editor of the journal *Moral Philosophy and Politics* (de Gruyter). For more information see https://homepage.uni-graz.at/en/lukas.meyer/ and https://online.uni-graz.at/kfu_online/wbforschungsportal.cbshowportal?pPersonNr=69399

Jörgen Ödalen is a senior lecturer in political science in the School of Business, Society, and Engineering at Mälardalen University, Sweden, and is an affiliate associate professor at the Department of Government, Uppsala University. His main area of research is contemporary political theory, with a particular focus on issues of climate-induced migration, territorial rights, and national and local self-determination. He has published papers in journals such as *Ethics, Policy and Environment*, *Lex Localis*, *Local Government Studies*, and *Urban Affairs Review*.

M. Feisal Rahman is a postdoctoral research associate in the Department of Geography at Durham University. Trained in environmental engineering, he works at the intersection of natural science, social science, and public policy.

His research to date has focused on climate change adaptation, assessing vulnerabilities to environmental changes, and enhancing the resilience of Asian deltas through transdisciplinarity, collaboration, and impact monitoring.

Kenneth Shockley is the Holmes Rolston III Endowed Chair in Environmental Ethics and Philosophy, is a professor of philosophy and affiliate faculty in the School of Global Environmental Sustainability at Colorado State University. His research interests are in climate ethics, ethical theory, the expression of environmental values in public policy, and the intersection between sustainable development, climate change, and environmental ethics.

Ivo Wallimann-Helmer is a professor of environmental humanities and director of the University of Fribourg Environmental Sciences and Humanities Institute. He specializes in ethical research on questions of justice and responsibility in environmental challenges. Key areas of his research are ethical conflicts in implementing climate measures and the normative implications of international policy decisions.

Felicia Wartiainen holds a BSc in political science and economics from Uppsala University and University of Illinois at Urbana-Champaign, specializing in international relations, sustainability studies, and migration. She is interested in the impacts of climate change on migration, humanitarian action, and sustainable development. She has worked as a global immigration and mobility consultant at EY Sweden and as a program associate at the United Nations Development Programme in Uganda on rule of law, access to justice, and human rights for refugees.

www.ingramcontent.com/pod-product-compliance
Lightning Source LLC
Chambersburg PA
CBHW020830020526
44115CB00029B/69